Essentials in Ophthalmology

Series Editor
Arun D. Singh
Cleveland, Ohio, USA

Essentials in Ophthalmology aims to promote the rapid and efficient transfer of medical research into clinical practice. It is published in four volumes per year. Covering new developments and innovations in all fields of clinical ophthalmology, it provides the clinician with a review and summary of recent research and its implications for clinical practice. Each volume is focused on a clinically relevant topic and explains how research results impact diagnostics, treatment options and procedures as well as patient management.

The reader-friendly volumes are highly structured with core messages, summaries, tables, diagrams and illustrations and are written by internationally well-known experts in the field. A volume editor supervises the authors in his/her field of expertise in order to ensure that each volume provides cutting-edge information most relevant and useful for clinical ophthalmologists. Contributions to the series are peer reviewed by an editorial board.

More information about this series at http://www.springer.com/series/5332

Chi-Chao Chan
Editor

Animal Models of Ophthalmic Diseases

 Springer

Editor
Chi-Chao Chan
National Eye Institute
National Institutes of Health
Bethesda
Maryland
USA

Zhongshan Ophthalmic Center
Sun Yat-sen University
Guangzhou
China

ISSN 1612-3212   ISSN 2196-890X (electronic)
Essentials in Ophthalmology
ISBN 978-3-319-79278-1   ISBN 978-3-319-19434-9 (eBook)
DOI 10.1007/978-3-319-19434-9

Springer Cham Heidelberg New York Dordrecht London
© Springer International Publishing Switzerland 2016
Softcover re-print of the Hardcover 1st edition 2016

Printed on acid-free paper

Springer International Publishing AG Switzerland is part of Springer Science+Business Media
(www.springer.com)

Foreword

The relatively recent development of therapeutics to treat and eventually cure disease has a trajectory built on the long historical relationship between basic (or "pure") science and "science for use" (clinical or translational science). This relationship is one that extends back at least to the time of Aristotle and has been driven by a dynamic tension based in part on different ways of thinking about science. One notion is that "pure" science and "science for use" are disparate categories and that basic science is primary and necessarily separated from its practical use by a long period of gestation (Stokes, 1997). However, there are many examples to show that this trajectory may not necessarily be unidirectional or of long duration; the present volume, "*Experimental Animal Models of Ophthalmic Diseases*" is one of them. What we see in these pages are the direct consequences of some very rapid recent advances in 21st century biology (eg, genomics) and their translation to clinical advances. This work is driven by the need to reduce what is a significant burden of visual system disease world-wide. The basic biology underlying these pre-clinical models involves fundamentally important contributions from genetics, epigenetics, inflammatory processes, aging, and neurodegeneration. Furthermore, this basic science is informed by a broad array of clinical investigations and clinical phenotypes, the latter often not completely expressed by animal models. The integration of this work with rapid technical and theoretical advances in engineering, physics, and informatics have allowed a much more rapid evaluation and development of clinical opportunities as recently envisioned in Precision (personalized) Medicine. All of these strands have been empowered by the exponential growth in computing capacity/speed and concomitant software design/development. The CRISPR/Cas9 system is a striking example of rapid biological engineering development. The translational use of this system has significantly accelerated over the last decade and was preceded, in the 1990's, by basic science advances involving genomic repeat elements in *E. coli* and other microbial systems (Hsu et al, Cell, 2014). CRISPR/Cas9 has transformed the use of transgenic as well as ES or iPS cell based disease models to mimic human disease phenotypes and provide therapeutic opportunities throughout the visual system (Zhong H, et al., Sci Rep., 2015) as suggested by one of the contributors. Cell based, human "disease in a dish" models also have considerable impact on the analysis of disease pathways, the identification of initiators of disease, and the utilization of high throughput screens for the discovery of small molecules and therapeutic drugs. The work described in the present volume and the rapid pace of 21st century scientific and clinical advances bodes very well for the treatment and cure of ocular diseases.

Sheldon Miller, PhD,
National Eye Institute

References

Hsu PD, Lander ES, Zhang F (2014) Development and applications of CRISPR-Cas9 for genome engineering. Cell 157(6):1262–78

Zhong H, Chen Y, Li Y, Chen R, Mardon G (2015) CRISPR-engineered mosaicism rapidlyreveals that loss of Kcnj13 function in mice mimics human disease phenotypes. Sci Rep 5:8366

Preface

When researching human diseases, models allow for a better understanding of the disease process without the added risk of harming an actual human. Like other forms of medical research, ophthalmology and vision research focuses on the investigation of disease pathogenesis and the discovery of novel therapies through *in vitro* and *in vivo* methodology. The *in vivo* experiments employ animal models including vertebrates (zebrafish, rodents, rabbits, and primates) and invertebrates (fruit flies and nematodes) for drug screening. Development of suitable experimental models is critical in identifying risk factors for disease, elucidating fundamental molecular mechanisms in disease progression, and providing guidance as to whether or not a particular treatment could be safe and effective for humans. This book is disease-oriented and presents different animal models used for common ocular diseases, including herpetic keratitis, cataract, glaucoma, age-related macular degeneration, diabetic retinopathy, uveitis, Graves' disease, and ocular tumors. In addition, world expert clinicians make critical comments on the clinical implications of each model.

Hendricks, Yun, Rowe, and Carroll compare some of the animal models of HSV-1 keratitis and their relation to human disease, and discuss some recent novel findings on the pathogenesis of HSV-1 keratitis in mice. Edward Holland states, "From a diagnostic and therapeutic perspective, HSV keratitis is one of the most challenging entities confronting the clinician." and "Animal models of HSV are critical to the understanding of the pathophysiology and efficacy of new treatment."

West-Mays describes the experimental animal models most commonly utilized for investigating the genetic and environmental risk factors known to contribute to cataract formation. The strengths and weaknesses of each of these models are highlighted, as well as many recent advances. Yizhi Liu commends: "Although animal models cannot fully represent the features of human cataracts, they are indispensable tools to explore the mechanism of cataractogenesis. Clinicians and researchers alike will find this chapter to be a guide for selecting proper cataract animal models according to different investigation purposes."

Johnson and Tomarev review the most important rodent models of glaucoma, highlight recent progress in the development of new glaucoma models, and explore the strengths and weaknesses of these models for studying human disease. John Morrison remarks "The rat and mouse both possess the essential cellular relationships that make the optic nerve head, the primary site of injury, unique. Coincident with the development of these models, modern cell biology methods are now available that allow us to study intra-ocular pressure-induced gene expression and protein changes in these small nerve heads." and that "They [these models] will ultimately lead to effective neuroprotective strategies for our most vulnerable patients."

Sennlaub discusses animal models of AMD risk factors that influence sub-retinal inflammation. He further outlines models as either "primary" due to genetic factors or "secondary" due to inflammatory responses. The use of these AMD models might help to understand the origin and role of the accumulation of subretinal mononuclear phagocytes in AMD; further, these models could help to identify drug targets able to inhibit the potentially pathogenic subretinal accumulation of these phagocytes or their neurotoxic and angiogenic mediators. Wai Wong affirms that the outlined concepts are helpful in putting the phenomenon of subretinal inflammation in the broad context of AMD pathogenesis. In fact, numerous AMD models have been developed both *in vitro* and *in vivo*. Animal AMD models have been reported in different species, mostly in the mouse.[1,2] Recently, the application of new technology to manipulate gene expression such as CRISPR/Cas9 genome editing has effectively caused retinal pigment epithelial defect in zebrafish.[3] Emily Chew states "Developing animal models to study human AMD is essential to our understanding of its pathogenesis and testing of potential therapies for this disease. Despite the lack of the macula in the mouse model and the inability to replicate all the characteristics of AMD in the mouse model, such as retinal pigment epithelial atrophy, the animal models will help us understand the basic processes involved in the pathobiology of AMD."

Chen and Stiff reveal the wide range of animal models of diabetic retinopathy (DR). They believe the rodent diabetes models will remain the most popular animal models for research into DR. Noemi Lois praises the authors, as they provide us with a thorough review of the experimental animal models of DR available, pointing out both their advantages and their shortcomings. She suggests "Basic scientists and clinicians should work together in the search for improved *in vivo* models and endpoints for research into DR with the final goal of improving the quality of life of people with DR."

Kielczewski and Caspi present the major animal models of non-infectious uveitis including the well-established and commonly used experimental autoimmune uveitis (EAU) as well as spontaneous genetic models and the humanized uveitis model. These models are important tools for vision researchers to unveil the complexities of ocular inflammation. On this, Robert Nussenblatt remarks "Clearly many experimental manipulations cannot be performed in patients, and animal models, which not perfect, provide the observer with the correct environment to observe and manipulate." However, he also warns "Correlating the observations from an animal model to the human can be a daunting task."

Chang focuses on mouse models of retinal degeneration and specifically, the use of the *rd10* (*Pde6b^rd10*) model as an example of how a retinitis pigmentosa (RP) model can be used to explore RP. The study of retinal physiology and pathophysiology in the *rd10* model has already led to a strong understanding of retinal development, maintenance, and function on a molecular, cellular, and tissue-specific level. Paul Sieving comments, "The chapter provides an extensive review of the many ways that biological knowledge has been developed regarding the causes and cellular consequences of the abnormal Pde6b protein in the rd10 mouse… this chapter reviews the diverse biological strategies that have been employed to rescue vision by slowing or ameliorating the pathophysiology… Despite this, attempts at human therapy for RP remain extremely limited to date."

Banga, Moshkelgosha, Berchner-Pfannschmidt, and Eckstein describe a mouse model of Graves' orbitopathy (GO) that recapitulates orbital inflammation and adipogenesis by genetic immunization with human TSHR ectodomain with close field electroporation. Rebecca Bahn thinks this model of GO, and future animal models evolving from it, will facilitate novel experimental approaches and new discoveries regarding GO pathogenesis. These *in vivo* studies will no doubt lead to randomized clinical trials and ultimately to more effective approaches to the care of patients with Graves' disease. Shivani Gupta and Raymond Douglas think although pathologic differences do exist between the animal model of Graves' orbitopathy and human disease, this model provides considerable advances to further elucidate the complex mechanisms that underlie GO.

Jager, Cao, Yang, Carita, Kalirai, van der Ent, de Waard, Cassoux, Aronow, and Coupland summarize various models used for ocular malignancies including melanoma, retinoblastoma, and lymphoma, highlighting the different species that can be used. Arun Singh notes that animal models are useful for not only basic research but also therapeutic testing and patient management in oncology.

In summary, this book, written by international authorities in the field of ophthalmology and vision research, provides a comprehensive review on the most highly relevant animal models for the most common ocular diseases. World expert clinician scientists give positive critiques on the use of these animal models based on their clinical experience.

Finally, I wish to thank the National Eye Institute where I was trained and have been working for over 33 years. I also need to thank all the authors; most of them are my colleagues and friends, who contributed the chapters on these common ocular disease models or made insightful comments from their clinical and scientific experience and knowledge. I would like to also thank Nicholas Popp for his editing and discussion. Lastly, I save my most profound thanks to my family and friends, whose love and encouragement helped me immensely through the process of this book.

Chi-Chao Chan, MD
National Eye Institute, National Institutes of Health, USA

Zhongshan Ophthalmic Center, Sun Yat-sen University, China

References

1. Ramkumar HL, Zhang J, Chan CC (2010) Retinal ultrastructure of murine models of dry age-related macular degeneration (AMD). Prog Ret Eye Res 29:169-190.
2. Fletcher EL, Jobling AI, Greferath U, Mills SA, Waugh M, Ho T, de Long RU, Phipps JA, Vessey KA (2014) Studying age-related macular degeneration using animal models. Optom Vis Sci 91: 878–886
3. Kotani H, Taimatsu K, Ohga R, Ota S, Kuwahara A (2015) Efficient multiple genome modifications induced by the crRNAs, tracrRNA and Cas9 protein complex in zebrafish. PLoS One 10(5):e0128319.

Contents

Contributors

Mary E. Aronow Clinical Branch, National Eye Institute, National Institutes of Health, Bethesda, MD, USA

Rebecca S. Bahn Mayo Clinic College of Medicine, Division of Endocrinology and Metabolism, Mayo Clinic, Rochester, MN, USA

J. Paul Banga The Rayne Institute, King's College London, London, UK

Universitäts-Augenklinik, Gruppe für Molekulare Ophthalmologie, Medizinisches Forschungzentrum, Universität Duisburg-Essen, Essen, Germany

Utta Berchner-Pfannschmidt Universitäts-Augenklinik, Gruppe für Molekulare Ophthalmologie, Medizinisches Forschungzentrum, Universität Duisburg-Essen, Essen, Germany

Scott Bowman Department of Pathology and Molecular Medicine, McMaster University, Hamilton, ON, Canada

Jinfeng Cao Department of Ophthalmology, Leiden University Medical Center, Leiden, The Netherlands

Kate L. Carroll Department of Ophthalmology, University of Pittsburgh School of Medicine, Pittsburgh, PA, USA

Department of Ophthalmology, and Graduate Training Program in Molecular Virology and Microbiology, University of Pittsburgh, Pittsburgh, PA, USA

Rachel R. Caspi Laboratory of Immunology, National Eye Institute, National Institutes of Health, Bethesda, MD, USA

Nathalie Cassoux Département d'oncologie chirurgicale, Institut and Laboratory of preclinical investigation, Department of Translational Research, Institut Curie, Paris, France

Chi-Chao Chan National Eye Institute, National Institutes of Health, Bethesda, MD, USA

Zhongshan Ophthalmic Center, Sun Yat-sen University, Guangzhou, China

Bo Chang The Jackson Laboratory, Bar Harbor, ME, USA

Mei Chen Centre for Experimental Medicine, Queen's University Belfast, Northern Ireland, UK

Emily Y. Chew Division of Epidemiology and Clinical Applications, National Eye Institute/National Institutes of Health, Clinical Trials Branch, Bethesda, MD, USA

Sarah E. Coupland Pathology, Department of Molecular and Clinical Cancer Medicine, Institute of Translational Research, University of Liverpool, Liverpool, UK

Nadine E. de Waard Department of Ophthalmology, Leiden University Medical Center, Leiden, The Netherlands

Didier Decaudin Laboratory of Preclinical Investigation, Translational Research Department, Institut Curie, Paris, France

Raymond S. Douglas Kellogg Eye Center, University of Michigan, Ann Arbor, MI, USA

Anja Eckstein Universitäts-Augenklinik, Gruppe für Molekulare Ophthalmologie, Medizinisches Forschungzentrum, Universität Duisburg-Essen, Essen, Germany

Charles E. Egwuagu Molecular Immunology Section, Laboratory of Immunology, National Eye Institute, National Institutes of Health (NIH), Bethesda, USA

Shivani Gupta Kellogg Eye Center, University of Michigan, Ann Arbor, MI, USA

Robert L. Hendricks Department of Ophthalmology, University of Pittsburgh School of Medicine, Pittsburgh, PA, USA

Department of Immunology, University of Pittsburgh School of Medicine, Pittsburgh, PA, USA

Department of Microbiology and Molecular Genetics, and Graduate Training Program in Molecular Virology and Microbiology, University of Pittsburgh School of Medicine, Pittsburgh, PA, USA

Edward J. Holland Cornea Services, Cincinnati Eye Institute, University of Cincinnati, Union, Kentucky, USA

Martine J. Jager Department of Ophthalmology, Leiden University Medical Center, Leiden, The Netherlands

Thomas V. Johnson Wilmer Eye Institute, Johns Hopkins Hospital, Baltimore, MD, USA

Helen Kalirai Pathology, Department of Molecular and Clinical Cancer Medicine, Institute of Translational Research, University of Liverpool, Liverpool, UK

Jennifer L. Kielczewski Laboratory of Immunology, National Eye Institute at National Institutes of Health, Bethesda, MD, USA

Yizhi Liu State Key Laboratory of Ophthalmology, Zhongshan Ophthalmic Center, Sun Yat-sen University, Guangzhou, China

Noemi Lois Department of Ophthalmology, Queen's University, Belfast, UK

Sheldon Miller National Eye Institute, National Institutes of Health, Bethesda, MD, USA

John C. Morrison Casey Eye Institute, Oregon Health and Science University, Portland, USA

Sajad Moshkelgosha The Rayne Institute, King's College London, London, UK

Universitäts-Augenklinik, Gruppe für Molekulare Ophthalmologie, Medizinisches Forschungzentrum, Universität Duisburg-Essen, Essen, Germany

Rohini M. Nair School of Medical Sciences, University of Hyderabad, Hyderabad, India

Robert Nussenblatt Laboratory of Immunology, National Eye Institute at National Institutes of Health, Bethesda, MD, USA

Alexander M. Rowe Department of Ophthalmology, University of Pittsburgh School of Medicine, Pittsburgh, PA, USA

Florian Sennlaub Inserm, U 968, UPMC University Paris 06, Paris, France

UMR_S 968, Institut de la Vision, UPMC University Paris 06, Paris, France

INSERM-DHOS CIC 503, Centre Hospitalier National d'Ophtalmologie des Quinze-Vingts, Paris, France

Paul A. Sieving National Eye Institute, National Institutes of Health, Bethesda, MD, USA

Arun D. Singh Department of Ophthalmic Oncology, Cole Eye Institute, Cleveland Clinic, Cleveland, OH, USA

Alan Stitt Centre for Experimental Medicine, Queen's University Belfast, Northern Ireland, UK

Stanislav I. Tomarev Section on Retinal Ganglion Cell Biology, Laboratory of Retinal Cell and Molecular Biology, National Eye Institute, National Institutes of Health, Bethesda, MD, USA

Wietske van der Ent Institute of Biology, Department of Pathology, Leiden University Medical Center, Leiden University, Leiden, The Netherlands

Judith West-Mays Department of Pathology and Molecular Medicine, McMaster University, Hamilton, ON, Canada

Wai T. Wong Unit on Neuron-Glia Interactions in Retinal Disease, National Eye Institute, Bethesda, MD, USA

Hua Yang Department of Ophthalmology, School of Medicine, Emory University, Atlanta, Georgia, USA

Hongmin Yun Department of Ophthalmology, University of Pittsburgh School of Medicine, Pittsburgh, PA, USA

Animal Models of Herpes Keratitis

Robert L. Hendricks, Hongmin Yun,
Alexander M. Rowe and Kate L. Carroll

1.1 Introduction to Herpes Keratitis

Herpes simplex virus type 1 (HSV-1) is a prevalent pathogen. Based on serologic testing, more than half of the human population in the USA (much higher in some developing countries) has been infected with HSV by age 70 (Xu et al. 2002), and recent studies of human cadavers suggest that the frequency of human TG harboring latent HSV-1 might be even higher than that suggested by serology (Hill et al. 2008). Herpes keratitis is an ocular condition that usually results from remittent-recurrent HSV-1 infections of the eye. In both humans and in a variety of animal models of herpes keratitis, primary HSV-1 infection of the eye or oral facial region leads to HSV-1 colonization of many sensory neurons

R. L. Hendricks (✉) · H. Yun · A. M. Rowe ·
K. L. Carroll
Department of Ophthalmology, University of Pittsburgh
School of Medicine, Pittsburgh, PA 15213, USA
e-mail: hendricksrr@upmc.edu

R. L. Hendricks
Department of Immunology, University of Pittsburgh
School of Medicine, Pittsburgh, PA 15213, USA

R. L. Hendricks
Department of Microbiology and Molecular Genetics,
and Graduate Training Program in Molecular Virology
and Microbiology, University of Pittsburgh School of
Medicine, Pittsburgh, PA 15213, USA

K. L. Carroll
Department of Ophthalmology, and Graduate Training Program in Molecular Virology and Microbiology,
University of Pittsburgh, Pittsburgh, PA, USA

in the trigeminal ganglion (TG). The virus has a special relationship with neurons in which the viral genome is retained as episomal DNA in the nucleus of the neuron in a quiescent state called latency. HSV-1 genomes can persist in the latent state in neurons throughout the lifetime of the person, or can periodically reactivate from latency, produce infectious virions, and transport these virions by axonal transport back to the corneal surface. These reactivation events can trigger recurrent herpes keratitis characterized by lesions that are restricted to the corneal epithelium and caused by virus replication and destruction of epithelial cells, or are focused in the stroma and are immunopathological in nature. The latter form is referred to as herpes stromal keratitis (HSK) or interstitial keratitis, and can lead to blindness due to progressive scarring of the cornea with recurrences. Recurrent HSK is the leading infectious cause of blindness in the USA and worldwide.

1.2 HSV-1 Latency

HSV-1 latency can be defined in molecular terms as retention of episomal viral DNA in the nucleus of a neuron with transcription of viral genes restricted to a family of latency-associated transcripts (LATs) that do not appear to be transcribed (Javier et al. 1988; Stevens et al. 1988). Alternatively, latency can be defined as retention of episomal HSV-1 DNA without production of infectious viral particles (Gardella et al. 1984). The latter definition permits low-level intermittent production of a limited array of viral lytic gene products without virus production during latency, which is supported by a growing body of data on latently infected mice (Kramer and Coen 1995; Feldman et al. 2002). These findings also raise the possibility that the immune system can respond to viral lytic gene products before the virus escapes latency, a concept that is also supported by experimental data.

Since reactivation from the latent state predisposes to recurrent herpes keratitis, a great deal of emphasis has been placed on studies of herpes latency and reactivation in animal models. These studies have demonstrated that ill-defined genetic characteristics of the virus determine the likelihood of a reactivation event. For instance, infection of rabbit corneas with the HSV-1 strains McKrae and Syn+ results in spontaneous reactivation from latency in the TG, whereas spontaneous reactivation does not occur following similar infections with strains such as KOS (Hill et al. 1987b). However, even with the more neurovirulent McKrae and Syn+ strains, spontaneous reactivation does not typically result in recurrent keratitis in rabbits. Moreover, even corneal infection with the more virulent McKrae and Syn+ strains does not typically result in spontaneous reactivation in mice. However, reactivation of the McKrae and Syn+ strains can lead to a high incidence of recurrent herpes keratitis following inductive signals to the cornea such as epinephrine in rabbits (Hill et al. 1987a, 1987b, 1987c; Berman and Hill 1985) and ultraviolet irradiation in mice (recently reviewed in Stuart and Keadle 2012).

Evidence in mice suggests that both viral- and host-encoded microRNAs (miRNAs), many encoded within the viral LATs, can repress HSV-1 lytic gene expression during latency (Umbach et al. 2008; Pan et al. 2014). A recent study suggested that siRNAs that regulate the neurovirulence factor γ34.5 may contribute to latency (Tang et al. 2008). Interestingly, it appears that the LAT region might function differently in latently infected mice and rabbits. HSV-1 deletion mutants lacking LAT show reduced lytic gene expression in mice (Leib et al. 1989; Garber et al. 1997), but dramatically increased lytic gene ex-

pression in rabbits during latency (Giordani et al. 2008). These species-specific differences in LAT function might maintain the viral genome in a more activated state in rabbits than in mice, accounting for spontaneous reactivation in rabbits. Recent evidence also suggests a role for epigenetic repression of viral gene expression during latency (Bloom et al. 2010). Using a mouse footpad infection model, the authors demonstrated that during latency the LAT locus is enriched for euchromatin marks associated with gene expression, while lytic genes are enriched for heterochromatin marks, consistent with lytic gene repression. Moreover, a growing body of evidence establishes that T cells (both CD8[+] and CD4[+] T cells) surround latently infected TG neurons in both mouse and human TG, and that CD8[+] T cells can block HSV-1 reactivation from a latent state both in vivo and in vitro in mice (reviewed in Rowe et al. 2013). Many of the CD8[+] T cells surrounding latently infected neurons in TG of both mice and humans exhibit an activation phenotype including expression of the lytic granule component granzyme B and production of interferon gamma (IFN-γ) (Knickelbein et al. 2008; Liu et al. 2000; Verjans et al. 2007; Theil et al. 2003). In mice CD8[+] T cells have been shown to form an immunologic synapse with neurons and to even release lytic granules into the synapse (Khanna et al. 2003; Knickelbein et al. 2008). Mice can use both IFN-γ and lytic granules to prevent HSV-1 reactivation without killing the infected neurons. These observations in latently infected mice call into question the once prevalent view that latent HSV-1 is ignored by the host immune system, and this has led many to hypothesize a more dynamic form of latency in which the virus is constantly trying to reactivate in a small number of neurons, but these attempts are counteracted in part by the surveillance of CD8[+] T cells and perhaps also CD4[+] T cells.

An interesting difference between mice and humans is the greatly increased sensitivity of transporters of antigen presentation (TAPs) to inhibition by the HSV-1 immediate early protein infected cell protein 47 (ICP47) in humans relative to mice. The important role of TAPs in loading peptides onto major histocompatibility complex (MHC) class I molecules for presentation to CD8[+] T cells suggests that HSV-1-infected cells might be less visible to CD8[+] T cells in humans than in mice. This effect might be more profound in latently infected neurons that express very low levels of MHC and viral proteins, and could account for the highly increased rate of spontaneous reactivation of HSV-1 in human TG relative to mouse TG. However, the fact that CD8[+] T cells surround latently infected neurons and show a similar activation phenotype in mice and human TG would seem to militate against this theory.

Identification of these immunologic and nonimmunologic mechanisms of regulating HSV-1 gene expression during latency informed current attempts to understand how these repressive mechanisms are regulated in animals. A variety of environmental and physiological stimuli (e.g., stress, exposure to UV irradiation, and immunosuppression) are associated with HSV-1 reactivation in both mice and humans. One common thread between these stimuli and HSV-1 reactivation is the association of both with suppression of T-cell function. For instance, the exposure of latently infected mice to restraint stress significantly inhibits the function of CD8[+] T cells that surround latently infected neurons in their TG and leads to HSV-1 reactivation from latency (Freeman et al. 2007). Moreover, persistent antigenic exposure of HSV-specific CD8[+] T cells in latently infected TG of mice can lead to a phenomenon called T-cell exhaustion, which is associated with a loss of T-cell function and a corresponding reduction in CD8[+] T-cell protection from reactivation (Jeon et al. 2013; St Leger et al. 2013). These findings suggest that bolstering the number and function of CD8[+] T cells in latently infected ganglia might represent an effective approach to reduce the rate of recurrent herpetic disease.

Given the apparently important role of HSV-specific CD8[+] T cells in controlling HSV-1 latency, one might predict that HSV-1 vaccines targeting CD8[+] T cells would be effective in preventing HSV-1 reactivation from latency. In fact, this approach has been advocated based on ex vivo studies with latently infected mouse TG

(Hoshino et al. 2007). However, a recent study provided the important caveat that adoptively transferred HSV-specific CD8[+] T cells (mimicking the effect of vaccination) were completely excluded from the TG after HSV-1 latency was established, even following induction of HSV-1 reactivation (Himmelein et al. 2011). These findings suggest that simply increasing the number and frequency of circulating HSV-specific CD8[+] T cells through vaccination might not be sufficient to increase the CD8[+] T cell population that is resident in latently infected TG. More heroic measures such as the use of neurotrophic viral vectors to express chemokines within the latently infected TG might be necessary adjunct to vaccination for this purpose.

1.3 HSV-1 Keratitis

Unfortunately, HSV-1 is able to escape the rather significant mechanisms for viral gene repression in many humans and frequently reactivates and is shed at the corneal surface. The consequences of such reactivation and shedding events can vary significantly in different individuals. In many individuals viral shedding into the tear film is completely asymptomatic (Kaufman et al. 2005). In others, viral reactivation and shedding can lead to recurrent lesions that are restricted to the corneal epithelium, referred to as infectious epithelial keratitis (Darougar et al. 1985). The lesions are caused by HSV-1 replication and destruction of epithelial cells, and for reasons that are not clear typically have a dendritic or geographic morphology. Still, in other individuals, HSV-1 reactivation and shedding can lead to recurrent bouts of HSK or interstitial keratitis. HSK is characterized by mild to severe inflammation in the corneal stroma with or without an overlying epithelial necrosis, referred to as necrotizing or immune HSK, respectively. This variability in HSV-1-induced corneal pathology appears to reflect the combined effect of genetic factors in the virus and the host.

These so-called "genetics squared" effects are difficult to study in outbred rabbits, but can be modeled to some extent in mice where different laboratory strains of HSV-1 induce varying levels of corneal disease in different inbred strains of mice. For instance, primary corneal infection of BALB/c mice with the KOS strain of HSV-1 induces epithelial lesions with subsequent development of necrotizing or immune HSK, whereas a similar infection of the corneas of C57BL/6 mice results only in epithelial disease without progression to HSK. In contrast, the RE strain of HSV-1 induces both epithelial lesions and progression to HSK following corneal infection of both BALB/c and C57BL/6 mice (Fenton et al. 2002). The complexity of the viral genome (containing approximately 86 open-reading frames) and the host genome has thus far defied rendering of a genetic explanation for this disease variability. However, recent advances in the RNA and DNA sequencing technology will likely increase the feasibility of such studies in the near future.

1.3.1   HSV-1 Epithelial Keratitis

HSV-1 epithelial keratitis has been studied extensively in rabbits and mice. In both the models the lesions have the characteristic dendritic or geographic morphology seen in humans. The duration of the epithelial lesions is somewhat longer in rabbits, making them a popular model system for studying the efficacy of antiviral drugs. Moreover, one can readily induce recurrent HSV-1 corneal epithelial lesions in rabbits by approaches such as corneal iontophoresis of epinephrine (Santos et al. 1987). Immune mechanisms that control HSV-1 replication in the corneal epithelium have been mainly studied in mouse primary corneal infection models. Since the resulting epithelial lesions heal within about 3–4 days, and the virus is typically cleared from the cornea by 7–8 days, it is not surprising that the innate immune response is primarily involved in viral clearance from the cornea in the mouse primary infection model. Mice lacking type-1 interferon (IFN-α and IFN-β) responses not only fail to clear the virus from the cornea but also experience disseminated lethal infections (Luker et al. 2003; Noisakran and Carr 2000; Carr and Noisakran 2002) (Hendricks et al. 1991). Mice lacking type

2 (IFN-γ) responses are less compromised, but do show delayed HSV-1 clearance from the cornea (Frank et al. 2012). Evidence suggests that viral clearance is primarily mediated by the combined action of the innate immune cells: natural killer (NK) cells, monocyte/macrophages, and cells expressing a γ/δ T-cell receptor (Ghiasi et al. 2000; Bukowski and Welsh 1986; Sciammas et al. 1997). Recent evidence demonstrates an important role of cornea-resident dendritic cells (DC) in regulating the migration of NK cells and monocytes into the HSV-1 corneal lesion (Frank et al. 2012). It should be noted, however, that most human corneal epithelial HSV-1 lesions result from reactivation of the latent virus and occur in the face of an established adaptive immune response. Therefore, the relative contribution of the innate- and adaptive immune system to HSV-1 clearance from the corneal epithelium needs to be examined in mouse or rabbit recurrent disease models.

1.3.2   Herpes Stromal Keratitis

HSK represents an immunopathological response that can occur in the absence of detectable replicating virus in both mice and humans. The disease is characterized by progressive leukocytic infiltration, opacity, and vascularization of the cornea that with recurrences can lead to progressive corneal scarring. From the seminal observation in mice that T cells are necessary for the development of HSK, mice have become by far the most popular model for studying the disease (Metcalf et al. 1979). In mice, HSK initiates around 7–8 days postinfection (dpi), after the detectable replicating virus has cleared from the cornea and enters a latent state in the TG. Following primary infection, HSK appears to be preferentially mediated by CD4[+] T cells in mice, although the infection of CD4[+] T-cell-deficient mice with a relatively high dose of HSV-1 leads to CD8[+] T-cell-mediated severe, but transient HSK that begins to resolve by around 20 dpi (Lepisto et al. 2006). HSK induced by primary infection of the mouse cornea is characterized by a diffuse inflammatory infiltrate in the corneal stroma consisting primarily of neutrophils with smaller, but significant populations of macrophages and CD4 T lymphocytes (Hendricks and Tumpey 1990). Their corneas become vascularized and edematous and the inflammation persists for months. Humans can develop similar severe HSK, but in most the disease is milder, more focal, and more transient. The difference might in part reflect the fact that HSK represents recurrent disease in most humans, but develops following primary infection in most mouse models. Indeed, when recurrent disease is induced in mice, HSK tends to be milder with focal stromal opacities (Miller et al. 1996). Histological examination of human corneas with HSK is usually restricted to corneal buttons removed during transplantation, which is typically performed only after inflammation subsides, so the leukocytic infiltrate is almost certainly not a reflection of that present during active HSK.

1.3.2.1 Neurotrophic Damage as a Major Component of HSK in Mice

Neurotrophic keratitis is corneal pathology resulting from impairment of trigeminal corneal innervation leading to a decrease or an absence of corneal sensation (Cruzat et al. 2011). The most frequent cause of corneal anesthesia is viral infection (especially herpes simplex and varicella zoster), followed by chemical burns, physical injuries, and corneal surgery. Corneal hypoesthesia (reduced sensation) is a hallmark of HSK in both mice and humans. The cornea is primarily innervated through the ophthalmic branch of the trigeminal nerve that is comprised of the frontal, lacrimal, and nasociliary nerves, with the latter providing sensory innervation of the cornea. The nasociliary nerves with a density of as many as 100 nerve endings per 0.01 mm^2 of corneal surface are responsible for the blink reflex that helps to maintain tear film protection of the corneal surface and prevents corneal inflammation associated with exposure keratitis. In vivo confocal microscopy reveals that the loss of corneal sensation in HSK correlates strongly with profound diminishment of the subbasal nerve plexus after HSV infection. A similar, though less profound retraction of the subbasal nerves can result from bacterial and fungal infections (Hamrah et al.

2010). Whether hypoesthesia associated with HSK is the direct effect of HSV-1 infection of nasociliary nerves, is caused by corneal inflammation, or results from a combination of both is unclear. The loss of sensitivity in human corneas with HSK rarely recovers, leading to the prevalent view that nerve loss is permanent.

Although the loss of corneal sensitivity and corneal nerves has been demonstrated in mouse models of HSK, until recently the contribution of nerve loss to corneal pathology was not evaluated. Mice lack consensual blink reflex (Yun et al. 2014), a reflex arc in humans that causes both eyes to blink when one cornea is irritated. Therefore, mice likely experience a more profound loss of blink reflex when corneal nerves are damaged by unilateral HSK. Infecting the corneas of BALB/c mice with 1×10^5 pfu of HSV-1 (RE or KOS strain) results in severe vascularization and dense corneal opacity that persists for at least 70 days after infection, and is associated with loss of corneal tactile sensation based on the loss of blink reflex (Yun et al. 2014). The loss of corneal blink reflex was associated with a dramatic reduction of corneal nerve fibers mainly involving the sensory endings and the plexus of nerve fibers extending from trunks in the corneal stroma into the epithelial layer of the cornea. Since the loss of corneal blink reflex can result in corneal desiccation and exposure keratopathy, the authors asked if the severe and chronic inflammation observed in the mouse HSK model could be ameliorated by protecting the cornea from desiccation by performing tarsorrhaphy. Tarsorrhaphy (stitching the eyelid closed) is a procedure commonly used in the clinic to protect corneas from desiccation. They observed that tarsorrhaphy could not only dramatically reduce inflammation in mouse corneas with HSK but also prevent the development of severe HSK when performed prior to its onset. Tarsorrhaphy did not prevent or repair corneal nerve damage or the associated loss of blink reflex, but did reduce inflammation associated with the corneal desiccation. It appears, therefore, that the severe and chronic inflammation associated with HSK in mice actually reflects exposure keratopathy associated with the inability to blink in response to corneal irritation. Thus, HSK in mice appears to reflect two distinct processes: a mild and potentially transient immunopathological response (presumably to HSV-1 antigens) resembling that seen in most humans; followed by a chronic severe immunopathological response to corneal desiccation.

As noted above, corneal nerve damage is thought to be irreversible in human patients with recurrent HSK. Indeed, nerve damage persists in mouse corneas with HSK for at least 70 dpi (longest follow-up in our studies). Interestingly, corneal infection of mice that are deficient in CD4$^+$ T cells results in severe, but transient HSK that resolves by around 20 dpi (Lepisto et al. 2006). The fact that severe HSK in mice is associated with loss of corneal nerves suggested that nerve regeneration might occur in the absence of CD4$^+$ T cells, which was shown to be true (Yun et al. 2014). Thus, the corneal nerve damage associated with HSK does appear to be reversible and inhibited by CD4$^+$ T cells. Characterization of how CD4$^+$ T cells inhibit nerve regeneration in mouse corneas with HSK might provide new therapeutic approaches to treating hypoesthesia and possibly reducing inflammation in people with HSK.

While most humans with HSK exhibit a milder and more transient inflammation, some exhibit severe, chronic HSK similar to that seen in most mouse models. Humans with unilateral HSK can exhibit varying degrees of nerve damage in their contralateral nondiseased eye (Hamrah et al. 2010). It is likely that initiation of consensual blink reflex in the contralateral eyes of these patients is somewhat impaired, resulting in desiccation that amplifies the severity and chronicity of virus-induced immunopathology in the infected eye. Also, the range of severity of human HSK might reflect the frequency of infected corneal nerves. When infecting mouse corneas, typically a large inoculum of HSV-1 (e.g., 1×10^5 pfu) is applied to a relatively small eye. Thus, a larger proportion of corneal nerves may be infected and targeted for damage in mice. Achieving a similar frequency of infected neurons in human corneas might require several episodes of HSV-1 reactivation, shedding at the corneal surface, and infection of additional nerves. Indeed, an association between neurotrophic damage and the frequency of HSV-1 recurrence has been noted in human corneas. Therefore, the severity and chronicity of HSK in humans might reflect the degree of nerve damage and the associated susceptibility to corneal desiccation.

1.4 Allografts on Corneas with HSK

When HSK recurrences result in progressive scarring and visual impairment, the only currently available therapy is to surgically replace the scarred cornea through full-thickness penetrating keratoplasty (PK) or deep anterior lamellar keratoplasty (DALK). However, corneal grafts that are performed to correct HSK-induced scarring are rejected with a much higher frequency than those performed to correct non-inflammatory damage. This is true even when transplantation is delayed until the cornea is clinically uninflamed. The findings from studies on both mouse and human models provide several possible explanations for the propensity for rejection of corneal transplants placed on these clinically uninflamed graft beds with a history of HSK.

First, we propose that the hypoesthesia associated with HSK in both mice and humans might contribute to the increased rate of rejection of grafts placed on HSK beds. Corneal nerves that emanate from the peripheral cornea are severed when the central cornea is excised during PK or DALK surgery, but an uninflamed corneal bed would retain corneal nerves and blink reflex. However, mice and humans with severe HSK can lose corneal BR in their entire cornea. Thus, neurotrophic inflammation resulting from the complete loss of corneal nerves might contribute to the high rate of transplant rejection in some HSK patients. Therapeutic approaches to reverse the nerve damage that persists in HSK corneas before or simultaneously with the cornea transplantation might increase the acceptance rate in cornea transplantation for HSK patients.

Also, one cannot assume that clinical examination can detect levels of corneal inflammation that can predispose to graft rejection. One retrospective study examined the corneal buttons obtained from HSK patients during transplantation

(Shtein et al. 2008). Although the recipient corneas had been declared clinically quiet by their physicians, their corneal buttons removed during grafting retained significant leukocytic infiltrates and the chemokines CCL2 and CXCL8 when examined histologically and biochemically. Patientswho had higher levels of CCL2 and CXCL8 were more likely to have graft failure or rejection when compared with patients who had truly quiet corneas. Thus, clinical examination is clearly not sufficient to detect levels of inflammation that can promote corneal graft rejection, and more sensitive methods are needed. Similar findings were obtained when BALB/c mice were infected with a dose of KOS HSV-1 that induced epithelial lesions, but no clinically detectable HSK (Divito and Hendricks 2008). These mice cleared their epithelial lesions by 3 dpi, and their corneas were then clinically uninflamed through at least 15 dpi. Nonetheless, their corneas retained significant levels of leukocytes and chemokines that are associated with HSK and corneal graft rejection, and corneal grafts placed on HSV-1 infected, but clinically uninflamed mouse corneas are rejected with a high frequency (our unpublished observation). Thus, mice with subclinical HSV-1 corneal infections might prove useful in defining mechanisms of rejection of corneal grafts placed on corneal beds with a history of HSK, but that lack clinical signs of inflammation at the time of grafting. This subclinical mouse corneal infection model might even have implications for corneal grafting in patients who exhibit subclinical shedding of HSV-1 in their tear film. If these corneas have subclinical inflammatory infiltrates similar to those observed in mice with subclinical infections, then they might be at increased risk for graft rejection. Treatment of such transplant recipients with antiviral drugs prior to transplantation might be indicated.

Compliance with Ethical Requirements
Robert L. Hendricks, Hongmin Yun, Alexander M. Rowe, Kate L. Carroll, and Edward J. Holland declare that they have no conflict of interest in the current work. No human studies were carried out by the authors for this article. All institutional and National guidelines for the care and use of laboratory animals were followed.

References

Berman EJ, Hill JM (1985). Spontaneous ocular shedding of HSV-1 in latently infected rabbits. Invest Ophthalmol Vis Sci26:587–90

Bloom DC, Giordani NV, Kwiatkowski DL (2010) Epigenetic regulation of latent HSV-1 gene expression. Biochim Biophys Acta 1799:246–56

Bukowski JF, Welsh RM (1986) The role of natural killer cells and interferon in resistance to acute infection of mice with herpes simplex virus type 1. J Immunol 136:3481–5

Car DJ, Noisakran S (2002) The antiviral efficacy of the murine alpha-1 interferon transgene against ocular herpes simplex virus type 1 requires the presence of CD4(+), alpha/beta T-cell receptor-positive T lymphocytes with the capacity to produce gamma interferon. J Virol 76:9398–406

Cruzat A, Witkin D, Baniasadi N, Zheng L, Ciolino JB, Jurkunas UV, Chodosh J, Pavan-Langston D, Dana R, Hamrah P (2011). Inflammation and the nervous system: the connection in the cornea in patients with infectious keratitis. Invest Ophthalmol Vis Sci 52:5136–43

Darougar S, Wishart MS, Viswalingam ND (1985) Epidemiological and clinical features of primary herpes simplex virus ocular infection. Br J Ophthalmol 69:2–6

Divito SJ, Hendricks RL (2008). Activated inflammatory infiltrate in HSV-1-infected corneas without herpes stromal keratitis. Invest Ophthalmol Vis Sci 49:1488–95

Feldman LT, Ellison AR, Voytek CC, Yang L, Krause P, Margolis TP (2002) Spontaneous molecular reactivation of herpes simplex virus type 1 latency in mice. Proc Natl Acad Sci U S A 99:978–83

Fenton RR, Molesworth-Kenyon S, Oakes JE, Lausch RN (2002). Linkage of IL-6 with neutrophil chemoattractant expression in virus-induced ocular inflammation. Invest Ophthalmol Vis Sci 43:737–43

Frank GM, Buela KA, Maker DM, Harvey SA, Hendricks RL (2012) Early responding dendritic cells direct the local NK response to control herpes simplex virus 1 infection within the cornea. J Immunol 188:1350–9

Freeman ML, Sheridan BS, Bonneau RH, Hendricks RL (2007) Psychological stress compromises CD8+ T-cell control of latent herpes simplex virus type 1 infections. J Immunol 179:322–8

Garber DA, Schaffer PA, Knipe DM (1997) A LAT-associated function reduces productive-cycle gene expression during acute infection of murine sensory neurons with herpes simplex virus type 1. J Virol 71:5885–93

Gardella T, Medveczky P, Sairenji T, Mulder C (1984) Detection of circular and linear herpesvirus DNA molecules in mammalian cells by gel electrophoresis. J Virol 50:248–54

Ghiasi H, Cai S, Perng GC, Nesburn AB, Wechsler SL (2000) The role of natural killer cells in protection of mice against death and corneal scarring following ocular HSV-1 infection. Antiviral Res 45:33–45

Giordani NV, Neuman DM, Kwiatkowski DL, Bhattacharje PS, McAnany PK, Hill JM, Bloom DC (2008) During herpes simplex virus type 1 infection of rabbits, the ability to express the latency-associated transcript increases latent-phase transcription of lytic genes. J Virol 82:6056–60

Hamrah P, Cruzat A, Dastjerdi MH, Zheng L, Shahatit BM, Bayhan HA, Dana R, Pavan-Langston D (2010) Corneal sensation and subbasal nerve alterations in patients with herpes simplex keratitis: an in vivo confocal microscopy study. Ophthalmology 117:1930–6

Hendricks RL, Tumpey TM (1990) Contribution of virus and immune factors to herpes simplex virus type 1 induced corneal pathology. Invest Ophthalmol Vis Sci 31:1929–39

Hendricks RL, Weber PC, Taylor JL, Koumbis A, Tumpey TM, Glorioso JC (1991) Endogenously produced interferon alpha protects mice from herpes simplex virus type 1 corneal disease. J Gen Virol 72:1601–10

Hill JM, Haruta Y, Rootman DS (1987a) Adrenergically induced recurrent HSV-1 corneal epithelial lesions. Curr Eye Res 6:1065–71

HillJM, RayfieldMA, HarutaY (1987b) Strain specificity of spontaneous and adrenergically induced HSV-1 ocular reactivation in latently infected rabbits. Curr Eye Res6:91–97

Hill JM, Shimomura Y, Dudley JB, Berman E, Haruta Y, Kwon BS, Maguire LJ (1987c). Timolol induces HSV-1 ocular shedding in the latently infected rabbit. Invest Ophthalmol Vis Sci 28:585–90

Hill JM, Bal MJ, Neuman DM, Azcuy AM, Bhattacharje PS, Bouhanik S, Clement C, Lukiw WJ, Foster TP, Kumar M, Kaufman HE, Thompson HW (2008) The high prevalence of herpes simplex virus type 1 DNA in human trigeminal ganglia is not a function of age or gender. J Virol 82:8230–4

Himmelein S, St Leger AJ, Knickelbein JE, Rowe A, Freeman ML, Hendricks RL (2011). Circulating herpes simplex type 1 (HSV-1)-specific CD8+ T cells do not access HSV-1 latently infected trigeminal ganglia. Herpesviridae 2:5

Holland EJ, Schwartz GS, Neff KD (2011) Herpes simplex keratitis. In: Krachmer JH, Mannis MJ, Holland EJ (eds) Cornea, 3rd edn. Mosby Year Book Publishers, St Louis

Hoshino Y, Pesnicak L, Cohen JI, Straus SE (2007) Rates of reactivation of latent herpes simplex virus from mouse trigeminal ganglia ex vivo correlate directly with viral load and inversely with number of infiltrating CD8+ T cells. J Virol 81:8157–64

Javier RT, Stevens JG, Dissette VB, Wagner EK (1988) A herpes simplex virus transcript abundant in latently infected neurons is dispensable for establishment of the latent state. Virology 166:254–7

Jeon S, St Leger AJ, Cherpes TL, Sheridan BS, Hendricks RL (2013) PD-L1/B7-H1 regulates the survival but not the function of CD8+ T cells in herpes simplex virus type 1 latently infected trigeminal ganglia. J Immunol 190:6277–86

Kaufman HE, Azcuy AM, Varnel ED, Sloop GD, Thompson HW, Hill JM (2005). HSV-1 DNA in tears and saliva of normal adults. Invest Ophthalmol Vis Sci 46:241–7

Khanna KM, Bonneau RH, Kinchington PR, Hendricks RL (2003) Herpes simplex virus glycoprotein B-specific memory CD8+ T cells are activated and retained in latently infected sensory ganglia and can regulate viral latency. Immunity 18:593–603

Knickelbein JE, Khanna KM, Ye MB, Baty CJ, Kinchington PR, Hendricks RL (2008) Noncytotoxic lytic granule-mediated CD8+ T cell inhibition of HSV-1 reactivation from neuronal latency. Science 322:268–71

Kramer MF, Coen DM (1995) Quantification of transcripts from the ICP4 and thymidine kinase genes in mouse ganglia latently infected with herpes simplex virus. J Virol 69:1389–99

Leib DA, Bogard CL, Kosz-Vnenchak M, Hicks KA, Coen DM, Knipe DM, Schaffer PA (1989) A deletion mutant of the latency-associated transcript of herpes simplex virus type 1 reactivates from the latent state with reduced frequency. J Virol 63:2893–900

Lepisto AJ, Frank GM, Xu M, Stuart PM, Hendricks RL (2006). CD8 T cells mediate transient herpes stromal keratitis in CD4-deficient mice. Invest Ophthalmol Vis Sci 47:3400–9

Liu T, Khanna KM, Chen X, Fink DJ, Hendricks RL (2000). CD8(+) T cells can block herpes simplex virus type 1 (HSV-1) reactivation from latency in sensory neurons. J Exp Med 191:1459–66

Luker GD, Prior JL, Song J, Pica CM, Leib DA (2003) Bioluminescence imaging reveals systemic dissemination of herpes simplex virus type 1 in the absence of interferon receptors. J Virol 77:11082–93

Metcalf JF, Hamilton DS, Reichert RW (1979) Herpetic keratitis in athymic (nude) mice. Infect Immun 26:1164–71

Miller JK, Laycock KA, Umphres JA, Hook KK, Stuart PM, Pepose JS (1996) A comparison of recurrent and primary herpes simplex keratitis in NIH inbred mice. Cornea 15:497–504

Noisakran SJ, Car DJ (2000) Therapeutic efficacy of DNA encoding IFN-alpha1 against corneal HSV-1 infection. Curr Eye Res 20:405–12

Pan D, Flores O, Umbach JL, Pesola JM, BentleyP, Rosato PC, Leib DA, Cullen BR, Coen DM (2014) A neuron-specific host microRNA targets herpes simplex virus-1 ICP0 expression and promotes latency. Cell Host Microbe15:446–456

Rowe AM, St Leger AJ, Jeon S, Dhaliwal DK, Knickelbein JE, Hendricks RL (2013). Herpes keratitis. Prog Retin Eye Res 32:88–101

Santos C, Briones O, Dawson CR (1987) Peripheral adrenergic stimulation and indomethacin in experimental ocular shedding of HSV. Curr Eye Res 6:111–8

Sciammas R, Kodukula P, Tang Q, Hendricks RL, Bluestone JA (1997) T cell receptor-g/d cells protect mice from herpes simplex virus type 1-induced lethal encephalitis. J Exp Med 185:1969–75

Shtein RM, Garcia DD, Musch DC, Elner VM (2008) HSV keratitis: histopathologic predictors of corneal allograft complications. Trans Am Ophthalmol Soc 106:161–8

St Leger AJ, Jeon S, Hendricks RL (2013) Broadening the repertoire of functional herpes simplex virus type 1-specific CD8+ T cells reduces viral reactivation from latency in sensory ganglia. J Immunol 191:2258–65

Stevens JG, Haar L, Porter DD, Cook ML, Wagner EK (1988) Prominence of the herpes simplex virus latency-associated transcript in trigeminal ganglia from seropositive humans. J Infect Dis 158:117–23

Stuart PM, Keadle TL (2012) Recurrent herpetic stromal keratitis in mice: a model for studying human HSK. Clin Dev Immunol 2012:728480

Tang S, Bertke AS, Patel A, Wang K, Cohen JI, Krause PR (2008) An acutely and latently expressed herpes simplex virus 2 viral microRNA inhibits expression of ICP34.5, a viral neurovirulence factor. Proc Natl Acad Sci U S A 105:10931–6

Theil D, Derfus T, Paripovic I, Herberger S, Meinl E, Schueler O, Strup M, Arbusow V, Brandt T (2003) Latent herpesvirus infection in human trigeminal ganglia causes chronic immune response. Am J Pathol 163:2179–84

Umbach JL, Kramer MF, Jurak I, Karnowski HW, Coen DM, Cullen BR (2008) MicroRNAs expressed by herpes simplex virus 1 during latent infection regulate viral mRNAs. Nature 454:780–3

Verjans GM, Hintzen RQ, van Dun JM, Poot A, Milikan JC, Laman JD, Langerak AW, Kinchington PR, Osterhaus AD (2007) Selective retention of herpes simplex virus-specific T cells in latently infected human trigeminal ganglia. Proc Natl Acad Sci U S A 104:3496–501

Waggoner-Fountain LA (2004) Grossman LB: herpes simplex virus. Pediatr Rev 25:86–93

Wald A, Corey L (2007) Persistence in the population: epidemiology, transmission. Human herpesviruses: biology, therapy, and immunoprophylaxis. Cambridge University Press, Cambridge

Xu F, Schillinger JA, Sternberg MR, Johnson RE, Le FK, Nahmias AJ, Markowitz LE (2002) Seroprevalence and coinfection with herpes simplex virus type 1 and type 2 in the United States, 1988–1994. J Infect Dis 185:1019–24

Yun H, Rowe AM, Lathrop KL, Harvey SA, Hendricks RL (2014) Reversible nerve damage and corneal pathology in murine herpes simplex stromal keratitis. J Virol 88:7870–80

Commentary

Edward J. Holland
e-mail: Eholland@holprovision.com
Cincinnati Eye Institute,
University of Cincinnati,
Union, KY, USA

The seroprevalence of herpes simplex viruses (HSV) worldwide is approximately 90 %. (Wald and Corey 2007). These pervasive and contagious pathogens are capable of causing both asymptomatic infection and active disease in a myriad of end organs. HSV can cause severe primary disease, particularly in children, neonates, and immunocompromised adults. In addition to oropharyngeal infections, encephalitis, meningitis, myelitis, erythema multiforme, hepatitis, and disseminated infection resulting in death, HSV can cause different kinds of ophthalmic diseases (Waggoner-Fountain 2004). Although research has resulted in a better understanding of the molecular biology and pathogenesis of HSV, infections caused by HSV remain a serious public health problem associated with significant morbidity and mortality. From a diagnostic and therapeutic perspective, HSV keratitis is one of the most challenging entities confronting the clinician. A variety of clinical manifestations of not only infectious keratitis but also immunologic disease can affect all levels of the cornea. Corneal disease includes infectious epithelial keratitis, neurotrophic keratopathy, immune stromal (interstitial) keratitis, necrotizing stromal keratitis, and endotheliitis (Holland et al. 2011).

Animal models of HSV are critical to the understanding of the pathophysiology and efficacy of new treatments. HSV-1 epithelial keratitis has been studied extensively in rabbits and mice. Animal models of HSV have allowed the study of the efficacy of antiviral medications both topical and systemic. In addition, these models provide a mechanism to gain insight into the pathophysiology of immune stromal (interstitial) keratitis, and a better understanding of the cause and treatments for neurotrophic keratopathy. Ultimately the animal models of HSV will lead to the most critical method of preventing morbidity and mortality from HSV—the development of vaccines to prevent primary infection and recurrent disease.

References

Holland EJ, Schwartz GS, Neff KD (2011) Herpes simplex keratitis. In: Krachmer JH, Mannis MJ, Holland EJ (eds) Cornea, 3rd edn. Mosby Year Book Publishers, St Louis

Waggoner-Fountain LA (2004) Grossman LB: herpes simplex virus. Pediatr Rev 25:86–93

Wald A, Corey L (2007) Persistence in the population: epidemiology, transmission. Human herpesviruses: biology, therapy, and immunoprophylaxis. Cambridge University Press, Cambridge

Animal Models of Cataracts

Judith West-Mays and Scott Bowman

2.1 Introduction

Cataract is an opacification of all or specific regions of the lens that results in obstruction of light and gradual loss of vision. It continues to be the leading cause of blindness worldwide despite the availability of effective surgery in the developed countries (WHO 1998, 2000, 2012). According to the World Health Organization (WHO), nearly 40 million people are blind worldwide, and of these, almost half of them are blind due to cataract (WHO 1998, 2000, 2012). A total of 82 % of the blind are more than 50 years of age, and thus the incidence of cataract is expected to increase further as the population ages. Primary cataract surgery is the most frequently performed surgical procedure in the developed world and remains the only cure, costing over US$ 3.5 billion each year in the USA alone (Allen 2007). This surgical intervention is not without any problems and can lead to a number of complications, the most common of which is secondary cataract, also known as posterior capsular opacification (PCO) (Bullimore and Bailey 1993; Kappelhof and Vrensen 1992; Wormstone 2002; West-Mays and Sheardown 2010).

The underlying causes of cataract are quite diverse with the majority of cataracts acquired after middle age (Zhang et al. 2012a). Congenital or juvenile cataracts, considered as early-onset cataracts, are less common but nonetheless have significant visual consequences. Congenital cataract can originate from genetic mutations restricted to the lens, accompany other ocular pathologies, or derive from a systemic genetic disorder, with estimates of incidence of 70, 15, and 15 %, respectively (Hejtmancik 2008; Reddy et al. 2004). Age-related cataracts can be classified into three main types, based on the location in which the opacity originates within the lens, such as the nucleus, cortex, or posterior pole, and each of these types has been correlated with different risk factors and populations based on geographical location (Zhang et al. 2012a). Epidemiological studies from around the world have identified a number of risk factors that are associated with age-related cataracts, including diabetes, sunlight (ultraviolet light) exposure, smoking, and steroid use (Zhang et al. 2012a; Beebe et al. 2010b). While these epidemiological investigations and genetic studies in humans have provided important clues toward the potential causes of cataracts, elucidation of the mechanisms underlying cataractogenesis has only been possible through the use of experimental animal models. Furthermore, because of the physiological consequence of cataract, loss of lens transparency, studying the lens in an in vivo and or ex vivo "whole lens" condition in animal models has been critical for understanding both the cause of cataract formation and testing potential preventative therapeutics. The goal of this chapter is to describe the

J. West-Mays (✉) · S. Bowman
Department of Pathology and Molecular Medicine,
McMaster University, HSC 1R10,
Hamilton, ON L8N3Z5, Canada
e-mail: westmayj@mcmaster.ca

© Springer International Publishing Switzerland 2016
C.-C. Chan (ed.), *Animal Models of Ophthalmic Diseases,* Essentials in Ophthalmology,

experimental animal models most commonly utilized for investigating the genetic and environmental risk factors known to contribute to cataract formation. We also attempt to highlight the strengths and weaknesses of each of these models as well as describe any recent advances.

2.2 Models of Congenital Cataract

A number of different animal models have been used for investigation of genetic-based cataracts (Table 2.1). Mouse models are most common as a mammalian system, due to short generation time, low cost of maintenance, and ability to manipulate the mouse genome with directed mutation. Rats are also used as their prevalence in laboratory breeding has led to the detection of spontaneous mutations. Zebra fish provide an excellent model for studying genes within lens development as they are small enough to be utilized in a large-scale format, and they develop quickly for multi-generational studies (Talbot and Hopkins 2000; Thisse and Zon 2002). Furthermore, the transparency and relatively large-sized eyes of zebra fish during development make them excellent models of ocular development and disease (Glass and Dahm 2004; Malicki 2000). Congenital cataracts isolated to the lens have so far been mapped to 39 loci in the human genome, 26 of which have been linked to mutations in specific genes that can be categorized by gene family. Nearly half of the known mutated genes in isolated congenital cataracts occur in the crystallin family, a quarter in the connexin family, and the remaining occur in cytoskeletal proteins and membrane proteins (Hejtmancik 2008). The review in this section will focus on the animal models utilized to understand these mutations, which are predominantly restricted to the lens.

2.2.1 Crystallins

The crystallin family of proteins makes up the bulk of the protein in the lens, and is the main component for the highly ordered structure that permits the refraction of light. The crystallins

Table 2.1 Experimental animal models of cataracts

Type of cataract	Animal model/ species	Key references cited in review
Congenital/Juvenile		
Transgenic		
Crystallins	Mouse	11, 12, 13, 17, 18, 20, 52, 53, 55, 173
	Rat	137, 138
	Zebrafish	45, 77
Connexins	Mouse	47, 122, 134, 161
Structural Proteins	Mouse	2, 16, 99, 125, 131, 132, 133
	Rat	158
Diabetic		
Induced	Rat	60, 69, 95, 103, 177
	Mouse	64, 74, 152
	Dog	67
Spontaneous	Rat	89, 97, 106, 127, 130
	Dog	174
UV-induced	Rat	35, 72, 80, 94, 139, 140, 151, 156
	Mouse	73, 93, 176
	Guinea pig	39, 118, 136, 179
	Calf	75
	Rabbit	40, 43, 179
Steroid induced	Chick embryo	57, 157
	Rat	81, 155
Oxygen induced	Rat	42, 76, 116, 178
	Guinea pig	4, 41, 50
	Rabbit	87, 144
Secondary/ posterior capsular opacification (PCO)	Rat	61, 79, 159
	Mouse	85
	Chick	153
	Bovine	128
	Rabbit	58, 59, 110
	Dog	19, 24, 114

can be divided into three classes: alpha, beta, and gamma, named in descending order by average molecular weight when they were first fractionated (Mörner 1894), each with their own subclass of isoforms. The α-class is coded by two genes *CRYAA* and *CRYAB*, and is notable for its chaperone function that protects against protein accumulation associated with temperature change.

Several mouse models exist involving mutation of the mouse genes *Cryaa* (αA) and *Cryab* (αB). Knockout mouse models were first used to determine the necessity of αA- and αB-crystallin in lens (and general ocular) development; αA-crystallin knockout mice exhibit microphthalmia and nuclear cataracts (Brady et al. 1997), and αA- and αB-crystallin double knockout mice have abnormal fiber cells and no lens sutures (Boyle et al. 2003). Additionally, αA-crystallin knockout mice have insoluble inclusion bodies inside the lens fiber cells containing αB-crystallin, suggesting that αA has a role in maintaining solubility of αB (Brady et al. 1997). αB-crystallin knockout mice, however, do not develop cataracts and have no significant ocular differences compared with wild type (Brady et al. 2001). Overall, the mouse knockout models of α-crystallin show that αA is integral in general lens development, as well as in maintaining the regular soluble structure of αB that permits lens transparency.

Mutations in the *CRYAA* gene were first detected at codon 116, and have been recorded in several different independent populations (Litt et al. 1998; Vanita et al. 2006; Beby et al. 2007; Hansen et al. 2007; Gu et al. 2008; Richter et al. 2008). Other sites of mutation have been detected in *CRYAA* in humans, all of which are isolated cataractous phenotypes (Pras et al. 2000; Mackay et al. 2003; Santhiya et al. 2006; Graw et al. 2006) or cataracts accompanied by microcornea (Hansen et al. 2007; Khan et al. 2007; Devi et al. 2008). Mouse models for four independent point mutations in the *Cryaa* gene have been described with one mutation in particular, a transition from arginine to cysteine at codon 54, observed in both mouse and humans (Xia et al. 2006b; Chang et al. 1996; Graw et al. 2001). Interestingly, while the mouse R54C mutation is a dominant mutation, human R54C mutations are inherited as recessive traits, suggesting different functional activities of αA-crystallins in the different species. The αA- and αB-crystallin proteins of the zebra fish have also been investigated, concluding that while the zebra fish αB-crystallin protein is truncated, with lower expression and activity than the human orthologue, the zebra fish and human αA-crystallin proteins are fundamentally similar in structure

and activity (Posner 2003). The zebra fish *cloche* mutant, which has been used as a model of hematopoiesis and vascular development, also exhibits deficiencies in αA-crystallin that coincided with cataract formation (Goishi et al. 2006). Decreased levels of αA-crystallin lead to insoluble γ-crystallin that prevent proper fiber cell differentiation, and result in opaque plaques in the lens. Recovery of γ-crystallin solubility upon overexpression of αA-crystallin supports the direct effect of αA chaperone activity of other crystallin proteins, and the necessity of αA-crystallin in lens fiber stability seen in mammalian lenses (Horwitz 2003).

Unlike α-crystallins, β- and γ-crystallins do not have chaperone function and instead confer more structural roles to the lens. The β- and γ-crystallins each contain two domains that fold into sheets to form a Greek key motif. Repeats of this motif allow dense packing of lens protein that minimizes light scattering. Human β-crystallin has two forms, more acidic and more basic (A and B, respectively), which are coded by six genes *CRYBA1,2,4* and *CRYBB1,2,3* that are suggested to be the result of gene duplication (Brakenhoff et al. 1992). The first β-crystallin mutation associated with a disease phenotype was the Philly mouse cataract, in which a long-known inherited cataract in mice (Kador et al. 1980; Zigler 1990) was linked to an altered β-crystallin protein (Nakamura et al. 1988) and finally to a mutation in the *Crybb2* gene (Chambers and Russell 1991). Normal βB-crystallin interacts with other β-crystallin protein in the lens, often in dimers, and the cataract formation caused by the *Crybb2* mutation results from the aggregation of mutant βB- with wild-type β-crystallins into insoluble, cataract-forming plaques.

Mutations in the genes encoding βA1/A3-crystallin proteins in mice and humans have also been recorded. *Cryba1* and its orthologues encode both the A1 and A3 protein variants of β-crystallin in mammals (Graw 1997), and wide-scale *N*-ethyl-*N*-nitrosourea (ENU)-mutagenesis in mice (Ehling et al. 1985) generated progressive cataracts from mutations in the *Cryba1* gene that were with a dominant inheritance pattern (Graw et al. 1999). In this study, the progression of the nuclear opacity followed the increasing expression of βA1/3-crystallin from E14.5 onward, indicating that the loss of lens transparency was directly proportional to the expression of the mutated protein. A colony of Sprague–Dawley rats with a nuclear cataractous phenotype has also been generated based on an inherited spontaneous mutation in the rat βA1/A3 gene dubbed Nuc1 (Sinha et al. 2005; Sinha et al. 2008). Although the genetic modifications of knockout mice are not available for replication in the laboratory rat, the rat has an eye that is larger and more complex than the mouse, permitting large-scale study of lenses more representative of human patients. The Nuc1 mutation is a ten base pair insertion into the Nuc1 βA1/A3 gene that causes a frameshift, adding ten new amino acids, interfering with the protein–protein interaction domain (Sinha et al.). Heterozygous Nuc1 mutants have congenital nuclear cataracts, whereas rats with homozygous Nuc1 mutations have prenatal lens rupture and abnormal vascular growth in the eye, including the retention of fetal vasculature and abnormal vascular patterning in the retina (Sinha et al. 2005; Zhang et al. 2005; Gehlbach et al. 2006). Functional Nuc1 is required for ocular development and necessary to maintain lens homeostasis (Parthasarathy et al. 2011).

Human γ-crystallin is encoded by six genes; *CRYGA* to *CRYGF*, encoding γA–γF-crystallin proteins. The γ-crystallins contain four Greek key motifs that fold to form higher order β-sheets, essential to maintain the transparent lens structure. Mouse models have been used to study the function of γ-crystallin through mutations. The first mutation in a γ-crystallin gene recorded was in the mouse *Cryge* gene called eye lens obsolescence (Elo) (Cartier et al. 1992). The Elo mutant is a dominant mutation that alters the reading frame of the fourth Greek key motif, thereby preventing the formation of the highly ordered crystallin structure. The mutant lens fails to develop properly before birth and causes postnatal microphthalmia. Since the Elo mutant, more than 20 independent mouse mutations in *Cryg* genes have been classified, spanning across all of the six *Cryg* genes; all are dominant mutations with isolated cataracts as the

only resulting phenotype (Graw 2009). Cataract formation in γ-crystallin mouse mutants involves the formation of amyloid fibrils in the nuclei of lens fiber cells, which disrupt normal transcriptional regulation, first shown in the γB^{nop}-mutant (Sandilands et al. 2002). These changes have also been observed in wild-type crystallins in lenses subjected to denaturing conditions (Meehan et al. 2004). It is hypothesized that the main role of α-crystallin chaperone protein is to ensure that the γ-crystallins remain stable, thereby preventing amyloid fibril formation (Hatters et al. 2001). Further mouse models of crystallin changes indicate accumulations of protein modifications that promote insolubility of the crystallins causing the cataract-forming aggregates develop over time in wild-type mice, similar to the phenotypes of the early-onset mutant models (Ueda et al. 2002). Modifications to the proteins over time, as well as changes to the lens environment over time, are the factors that contribute to the development of senile cataracts (Wu et al. 2014). Finally, the zebra fish animal model has also been used for determining the molecular mechanism by which γ-crystallin mutations cause cataracts. An inherited γ-crystallin mutation was found in a three generation Chinese family, Gly129Cys, and the mutation was generated in the zebra fish to observe changes in lens development (Li et al. 2012).

2.2.2 Connexins

Connexins are gap junction proteins that facilitate exchange of small molecules between cells, particularly important in maintaining homeostasis in avascular tissues such as the lens in which lens transparency relies on a balanced osmotic and metabolic environment (Goodenough 1992). Three connexin genes expressed in the mammalian lens have been identified, Cx43, 46, and 50, and are encoded by the genes *Gja1*, *Gja3,* and *Gja8*, respectively. Mutations found in the genes *Gjp3* (encoding Cx46 or α$_3$) and *Gjp8* (encoding Cx50 or α$_8$) have been linked to cataracts with dominant and recessive inheritance. Homozygous Cx46-null mice developed nuclear cataracts that formed due to aggregation of lens protein, facilitated by the degradation of crystallins (Gong et al. 1997). This mouse model showed normal early lens development, but with small opacities forming 2–3 weeks after birth that continued into full nuclear cataracts by 2 months, with 100% penetrance of cataracts reported, indicating Cx46 is required for proper lens function (Gong et al. 1997). Interestingly, the severity of the cataracts on the Cx46-null model varied depending on the breeding background of the mouse, with milder cataracts observed in C57B16 mice when compared with mice bred from the 129 strain (Gong et al. 1999). The insoluble aggregates formed in the nuclear cataracts show disulfide cross-linking between the proteins, resembling human senile cataracts (Garner et al. 1981).

Mutations in the human Cx50 gene were first linked to cataracts by identifying a missense mutation in a family with inherited cataracts (Shiels et al. 1998). Since then, more than 15 mutations have been observed in humans at various codons linked to cataracts. In Cx50 homozygous knockout mice, microphthalmia is observed after 2 weeks (White et al. 1998; Rong et al. 2002). Additionally, small opacities in the lenses were observed at 2 weeks, which continued through 6 months exhibiting clustered nuclear opacities, reflecting the zonular pulverulent phenotype seen in the first observed Cx50 human mutation (Shiels et al. 1998). The Cx50 knockout mouse model suggests not only that Cx50 is required for proper fiber cell protein solubility, but also that Cx50 plays a more general role in lens development, which is known to be correlated with ocular growth (Coulombre and Coulombre 1964). White et al. replaced Cx50 function by knocking-in Cx46 and found that while the lens development defects of Cx50-null lenses were not corrected by Cx46-induced expression, the cataract phenotype of Cx50-null mice was prevented (White 2002; Martinez-Wittinghan et al. 2003). The Cx46-replacement mouse model showed that while Cx50 is required for normal lens development, recovery of nonspecific connexin signaling is enough to maintain lens homeostasis lost by knocking out Cx50 and prevent cataract formation. Point mutations have

also been detected in the mouse Cx50 gene. Of note, while the heterozygous knockout of Cx50 did not produce a cataract phenotype, individual point mutations of the Cx50 gene in mice lead to dominant cataracts in mice (Mathias et al. 2010). It is hypothesized that in the formation of the six-membered connexin hemichannels, mutated connexins act as dominant negative inhibitors of wild-type connexins at gap junctions, causing the disruption in homeostasis (Xia et al. 2006a; DeRosa et al. 2007; DeRosa et al. 2009; Berthoud and Beyer 2009; Xia et al. 2012; Rubinos et al. 2014).

2.2.3 Cytoskeletal and Membrane Proteins

Major intrinsic protein (MIP), also known as aquaporin 0, is a voltage-dependent water channel and the most abundant membrane-associated protein in lens fiber cells (Bloemendal et al. 1972; Gorin et al. 1984). Models of disruption in the mouse gene *Mip* determine that the cataractous phenotype of *Mip* mutant mice is due to disruption in the fiber cell differentiation, and subsequent disruption protein solubility in the crystallin lens (Shiels and Bassnett 1996; Sidjanin et al. 2001; Okamura et al. 2003; Shiels et al. 2001; Al-Ghoul et al. 2003). The mouse Hfi mutation presents as full total lens opacity in homozygous mice, because of the toxic function of mutated MIP protein. Recently, a rat spontaneous mutation was observed; the KFRS rat model was found to have a spontaneous frame shift mutation in the kfrs4 gene, which inactivates MIP causing recessive cataracts (Watanabe et al. 2012).

Mutations in structural proteins of lens cells have also been linked to cataracts. Vimentin is a type-III filament that is found in both lens epithelial cell (LEC) and fiber cell and has been shown to interact with α-crystallin (Ramaekers et al. 1980; FitzGerald 2009; Djabali et al. 1997; Nicholl and Quinlan 1994). Mice have been used successfully to determine the effect of vimentin expression on lens opacity; vimentin knockout mice do not present with cataracts (Colucci-Guyon et al. 1994), however mice in which a vimentin transgene was used to increase endogenous lens levels of vimentin tenfold developed cataracts (Capetanaki et al. 1989). In the vimentin overexpressing mice, normal fiber cell differentiation was disrupted and newly formed fiber cells did not elongate or undergo proper denucleation. Similarly, a mouse knockout model of HSF-4, a chaperone factor that represses vimentin overexpression, showed major disruption in lens size and differentiation, resulting in cataracts resembling the vimentin transgenic mice (Mou et al. 2010). Another structural protein that has been linked to cataract formation is beaded filament structural protein 2. This intermediate filament is present in differentiated lens fiber cells, interacting with α-crystallins and other intermediate filaments, and knockout mice develop opacification (Alizadeh et al. 2002; Sandilands et al. 2003).

2.3 Models of Age-Related Cataracts

2.3.1 Diabetic Cataract

Diabetes mellitus affects approximately 285 million people worldwide and the International Diabetes Federation reports that this is likely to increase to 439 million by 2030 (Pollreisz and Schmidt-Erfurth 2010). Premature cataract formation is one of the earliest secondary complications of diabetes and is considered a major cause of visual impairment in diabetic patients. While these cataracts can be successfully treated with surgery, diabetic patients are more susceptible to complications (Stanga et al. 1999). A complex interaction between genetic factors and environmental factors contributes to the pathogenesis of diabetes in humans; however, this has been difficult to verify (Shiels and Hejtmancik 2007). The use of experimental animal models has therefore been essential in further resolving these issues and in clarifying the pathogenetic mechanism(s) of diabetic cataract. Experimental diabetes mellitus has been induced using a number of different methods and in multiple animal models (Table 2.1). Most common methods for inducing diabetes involve pharmacological agents such as alloxan or streptozotocin (STZ), which selectively damage the insulin-producing cells in the pancreas resulting in hyperglycemia. Hyperglycemia is the common contributing factor to cataract in both type-1 and type-2 diabetes (Pollreisz and Schmidt-Erfurth 2010). Using this method, rodents, and in particular rats, have been the most commonly employed experimental model for investigating diabetic cataract. Although larger than mice, rats are still easy to handle and the cost of maintenance is lower than larger mammals. Moreover, diabetes can be induced quickly and effectively in the rat with a single dose of STZ or by galactose feeding resulting in features reminiscent of type-1 diabetes observed in humans. In the STZ model, a substantial (approximately 30 nmol/g lens) accumulation of sorbitol and galactitol occurs in the lens, resulting in the development of "fast" sugar lens opacities (Obrosova et al. 2010). The high level of sorbitol is thought to initiate osmotic stress in the lens that in turn results in infusion of fluid leading to lens fiber cell swelling and liquefaction. While the specific mechanism(s) by which hyperglycemia causes human cataracts has not been resolved, use of the diabetic rat model has helped to understand some potential, including activation of aldose reductase (AR), an enzyme that catalyzes the reduction of glucose to sorbitol through the polyol pathway, oxidative stress, and the generation of reactive oxygen species (ROS) and nonenzymatic glycation/glycoxidation. More recently, studies using excised rat lenses have also shown that when cultured with high glucose media lenses respond to the osmotic stress by altering the expression of key growth factors such as b-fibroblast growth factor (FGF) and transforming growth factor (TGF) involved in cell signaling, and these may contribute to the cataract formation (Zhang et al. 2012b).

In addition to the STZ and galactosemic rats, there also exist a number of rat strains that exhibit spontaneous onset diabetes and associated cataracts (Table 2.1). These include the Goto-Kakizaki (Ostenson and Efendic 2007), Zucker Diabetic Fatty (Schmidt et al. 2003), Otsuka Long-Evans Tokushima Fatty (OLETF) (Matsuura et al. 2005), Wistar Bonn/Kobori (WBN/Kob) rats (Mori et al. 1992) and the spontaneously diabetic tori rat (SDT) (Sasase et al. 2013).

In the WBN/Kob rat, plasma glucose levels are elevated at 10 months of age in males only and the cataracts are observed by approximately 12 months. The OLETF rat is characterized by late-onset glycemia, mild obesity, and male inheritance. The AR and sorbitol levels in the lens of this rat strain spike at 40 weeks of age with noticeable cataractous changes at 60 weeks. The SDT is an inbred strain of Sprague–Dawley rat that has gained recent attention since it is the only spontaneous strain that exhibits severe diabetic retinopathy (Sasase et al. 2013). The model also exhibits diabetic cataracts: In the male SDT rats, cataracts are fully penetrant by 40 weeks of age. Opacities are first evident at the posterior pole of the lens and then nuclear sclerosis progresses and the cortex becomes highly opacified. The cataractous lens exhibits features such as swollen lens fibers, liquefaction, vacuolation, abnormal configuration, and the formation of Morgagnian droplets and eventually capsular rupture. Importantly, these complications can be prevented in the SDT rat by normalizing blood glucose with insulin treatment or pancreas transplantation. In addition, the histopathological changes of the lens were preceded by an increase in lens sorbitol content, further suggesting that the cataracts in the SDT rats are a result of sustained hyperglycemia (Sasase et al. 2013).

Mice are used in many in vivo studies since they are small in size resulting in easier handling and less cost for housing. In addition, the availability of numerous transgenic and knockout strains enables investigators to study the role of a particular gene in disease. However, despite severe hyperglycemia (induced or spontaneous), diabetic mice do not develop cataracts and the levels of AR in the lens remain low (Varma and Kinoshita 1974). In addition, STZ administration in mice with a single dose, as carried out in the rat, causes lethality typically by 1 month of age, too short of a survival time for the appearance of cataracts. Thus, the mouse has not been a readily used model for diabetic cataract and transgenic models that are used to study other diabetic complications cannot be used to study diabetic cataract. Some attempts have been made, however, with limited success, to create a mouse model

of diabetic cataract, by using an STZ multidose method (Hegde et al. 2003). Additionally, when AR is introduced into the lenses of transgenic mice, sugar cataracts do form under both diabetic and galactosemic conditions, rendering this a potential model (Lee et al. 1995).

Animal models have also been employed to investigate a number of anti-cataract agents, including pharmaceuticals that could target the various mechanisms underlying diabetic cataract formation, some of which are readily available over the counter medicines or health food supplements. Of these, aldose reductase inhibitors (ARIs) have been explored in great depth, mainly using diabetic models in rat (Kador et al. 2001; Moghaddam et al. 2005; Halder et al. 2003). A number of plant-derived, natural products known to inhibit AR activity have been explored as well as intrinsic ARI-containing extracts obtained from human kidney and bovine lenses (Kador et al. 2001). These studies provide evidence that ARIs may prevent and delay diabetic cataract formation.

Experimental induction of diabetes has also been carried out in larger mammals such as the rabbit and the dog (Table 2.1). In the dog, cataract, lens rupture, and lens-induced uveitis are well-known consequences of spontaneous and induced diabetes (Zeiss 2013). Importantly, the topical ARI drug, Kinostat has shown promise in its ability to reverse early sugar cataract development in dogs (Kador et al. 2006).

2.3.2 UV-Induced Cataracts

Epidemiological research suggests a strong association between exposure to ultraviolet radiation-B (UVB) and cortical cataract formation in humans (Zigman et al. 1979; Taylor 1980; Delcourt et al. 2000; McCarty and Taylor 2002). The main UVR source is the sun and because of the fact that the ozone is being depleted, chances of exposure are greater today. While the cornea is the major absorber of UV radiation reaching the eye, UVA and the longest wavelengths of UVB radiation are absorbed by the lens and have been shown to damage lens proteins both in vivo and in vitro (Moreau and King 2012). In addition, an increase in oxidizing species may occur as a result of UV irradiation. The experimental animal models used to study the effect of UVR exposure on the lens have been quite varied since the lenses of many species are susceptible to UVR-induced cataractogenesis (Table 2.1). Most of the studies involve smaller species such as rats, mice, guinea pigs, and fish. However, rabbits and pigs have also been employed. The majority of these experiments involve anesthetizing the animals, dilating the pupils with mydriatic eye drops, and exposing them to either UVA or UVB of varying dosages, with the most damaging wavelengths located around 300 nm (Pitts 1978; Merriam et al. 2000). In other cases, however, animals are euthanized, lenses removed, and irradiated in vitro.

The mouse has been used in many UV-induced cataract models in order to elucidate the genetic and biochemical factors at play. For example, transgenic and knockout mice have been used to study the specific role of proteins in protecting the lenses against UV-induced cataracts (Lassen et al. 2007) (Meyer et al. 2009). As an example, UVB-induced cataract formation was found to be accelerated in mice with single and double knockout of the genes encoding the AL-DH1A1 and ALDH3A1, establishing the importance of these two enzymes in the mechanism of cataract formation (Lassen et al. 2007). Furthermore, in these KO mice decreased proteasomal activity, and increased oxidative damage of proteins, was observed in the double knockout mice as compared with wild-type animals. These types of studies can help to further understand how proteins in the lens and cornea can provide added safeguards to prevent UV-induced cataract formation and maintain lens clarity. Recent work using the mouse model has also revealed that older animals are more susceptible to UV-induced cataract formation than young mice (Zhang et al. 2012a). This may be related to the elevated expression of the thiol damage repair enzymes thioltransferase and thioredoxin found only in young lenses (Zhang et al. 2012a).

The albino Sprague–Dawley rat has also been a common model over the years with an extensive database on UVB-induced cataract

(Galichanin et al. 2010; Soderberg 1990; Soderberg et al. 2002; Wang et al. 2010). Much of these data, along with other species, have revealed important mechanisms underlying UVR-induced cataract, particularly for UVB and the development of cortical cataract. For example, UVB can cause direct DNA damage and generate ROS, singlet oxygen, superoxide, hydrogen peroxide, and hydroxyl radicals leading to oxidative stress and apoptosis of LECs (Michael et al. 1998; Wang et al. 2010). UVA and UVB have also been shown to directly damage lens proteins both in vivo and in vitro (Giblin et al. 2002; Simpanya et al. 2008). Most recent work using the UVB-induced rat model has focused on determining potential anti-cataract/protective agents. For example, caffeine has been shown to have in vitro and in vivo protective effects against UVB irradiation in rats (Kronschlager et al. 2013; Varma et al. 2008). The lens contains a high concentration of reduced glutathione (GSH), which acts to eliminate oxidants and protect against exogenous and endogenous ROS. Caffeine is thought to protect against UVR-induced cataract by scavenging ROS (Giblin 2000; Lou 2003; Spector et al. 1985; Varma et al. 2008).

Like humans, lenses of diurnal species, such as duck, frog, guinea pig, rabbit, and squirrel, contain high levels of UVA chromophores, in contrast to lenses of nocturnal species, such as rat and cat (Wood and Truscott 1994; Zigler and Rao 1991). Thus, these species have been used for studying UVA-induced early nuclear cataract (Giblin et al. 2002). In particular, studies on guinea pig provide in vivo evidence that link UVA-induced protein aggregation with an increased level of nuclear light scattering. The guinea pig lens is thought to mimic the aging human lens because it possesses a high level of a protein-bound toxic chromophore—not protein-bound kynurenine as in the aging human lens (Parker et al. 2004), but protein-bound NADPH in the form of zeta crystallin, which makes up 10 % of total lens protein in the guinea pig (Rao and Zigler 1990; Zigler and Rao 1991). Presumably, this along with trace amounts of molecular oxygen leads to UVA-induced generation of ROS. A similar process of UVA-induced protein aggregation may take place in the older human lens nucleus, possibly accelerating the formation of human nuclear cataract. Finally, UVA exposure has also been shown to induce cataracts in calves (Lee et al. 1999) and in vitro cataract development in pig lenses (Oriowo et al. 2001).

The rabbit lens also contains high levels of the UVA-absorbing pyridine nucleotide NADH, both free and bound to l-crystallin (Giblin and Reddy 1980; Zigler and Rao 1991). Many of the rabbit studies investigate the preventive effect of contact lenses containing UV-blocking materials. These findings show that contact lenses are beneficial in protecting ocular tissues of the rabbit against harmful effects of UV light, including photokeratitis and anterior subcapsular cataract (Giblin et al. 2012).

2.3.3 Steroid-Induced Cataracts

The use of steroid therapy in recent years has contributed to an increase in the incidence of cataracts (Gupta and Wagner 2009; Watanabe et al. 2000). Administration of synthetic glucocorticoids (GCs), or corticosteroids, as anti-inflammatory or immunosuppressive agents have been used in the treatment of many diseases including rheumatoid arthritis and asthma as well as in organ transplants and chemotherapy. The cataracts formed by steroid use typically occur in the posterior region of the lens, beneath the lens capsule and are thus referred to as posterior capsular cataracts (PSC) (Eshaghian and Streeten 1980). PSCs exhibit swollen fiber-like cells in the posterior pole of the lens, which have retained nuclei. The association between steroid use and the formation of cataracts in humans was made in the 1960s following the examination of 72 patients being treated for rheumatoid arthritis (Black et al. 1960). In all, 42 % of the patients developed cataracts whereas none of the patients in the control group did. Despite knowing GCs induce PSC, either through topical or systemic treatment, it remains unknown whether the effect is due to direct action on lens cells by binding to glucocorticoid receptors (GRs) or by indirect mechanisms.

A small number of animal models have been developed to study the underlying causes of steroid-induced cataracts (Table 2.1). The chick embryo has been used to study the response of the lens to GC. GRs have not been identified in the chick, and thus, GCs are thought to act on lens cells indirectly. It has been proposed that they do so by binding to GRs in the liver resulting in an increase in blood lipid peroxides that travel to the aqueous humor and deplete the lens of glutathione (Gupta and Wagner 2009; Watanabe et al. 2000). This in turn is thought to result in an increase in oxidative stress and formation of cataracts (Gupta and Wagner 2009; Watanabe et al. 2000). These findings are somewhat paradoxical, however, since treatment of chick lenses alone with GC can also develop cataracts, suggesting that additional mechanisms are at play in GC-induced cataract formation.

More recently, the mammalian lens has been shown to express GR, suggesting that direct activation by GC may also cause PSC formation. For example, the development of PSC in rabbits and rats treated with GC has been reported. Rats have been used more extensively as a mammalian model for exposure to high-dose steroids and this resulted in altered expression of the cell adhesion molecules, E- and N-cadherin, suggesting this may contribute to the disruption in fiber cell differentiation observed in PSC (Lyu et al. 2003). To further explore the mechanism of response of lens cells to GC rat ex vivo lens explants have also been used. This study demonstrated that treatment of explants with the GC, dexamethasone, along with the growth factor, fibroblast growth factor 2 (typically found in the aqueous and vitreous humors), resulted in an increase in cell proliferation and coverage on the capsule, occurrences observed in PSC formation (Wang et al. 2013). Future research has been proposed using the rat model to explore potential signaling pathways induced by GC including ERK1/2, known to be involved in fiber cell differentiation (Wang et al. 2013). Some caution has been made, however, that some differences in steroid-induced signaling may occur in rats as compared with the human lens because of the fact that other receptors are present in the human lens such as the GR receptor (Gupta and Wagner 2009). Thus, it was suggested that findings in the rat should be verified with using human lenses (Gupta and Wagner 2009).

2.3.4   Oxygen and Nuclear Cataracts

Typically, the lens resides in a hypoxic environment and increased exposure to oxygen is thought to be a significant cause of age-related nuclear cataracts (Beebe et al. 2010a). For example, it has been shown that individuals who received hyperbaric oxygen (HBO) therapy for more than 1 year developed nuclear cataracts (Beebe et al. 2010a; Palmquist et al. 1984). Additional evidence to support this comes from patients undergoing vitrectomy, which involves the destruction and removal of vitreous humor, as is typically done during retinal surgery. Following vitreous removal the lens is exposed to high levels of oxygen and these patients exhibit a high incidence (60–95%) of nuclear cataract formation (Cherfan et al. 1991; de Bustros et al. 1988; Thompson et al. 1995). The intact biochemical nature and gel structure of the vitreous are thought to protect the lens from oxygen and nuclear cataract formation. With age the vitreous undergoes collapse, referred to as vitreous syneresis, and loses its ability to maintain a low oxygen tension around the lens (Beebe et al. 2010a).

Animal models have been used to both measure oxygen levels in the vitreous and determine the mechanism of how high oxygen levels cause cataract formation (Table 2.1). Initially, studies using animal models confirmed the effect of high oxygen exposure on nuclear opacification. The main model for this has been the guinea pig, which when exposed to multiple treatments of HBO over extended periods of time exhibited lenses with increased nuclear scatter, along with protein insolubilization and increased levels of oxidized glutathione in the lens nucleus (Giblin et al. 1995; Bantseev et al. 2004; Gosselin et al. 2007). Studies using both cats and rats have shown the importance of the vitreous in protecting the lens from increased oxygen levels. For example, when the cat vitreous is disrupted enzymatically

increased oxygen levels occur. Rats exposed to HBO along with a loss of the vitreous resulted in decreased levels of protective enzymes in the lens and an increased level of glutathione protein mixed disulfides, unlike those animals in which the vitreous was left intact (Quiram et al. 2007; Giblin et al. 2009; Li et al. 2013). Further research using rats has shown that when hyperoxic conditions induce nuclear cataracts this is associated with mitochondrial DNA damage (Zhang et al. 2010). Some investigators have developed synthetic gels to replace the vitreous after vitrectomy and these studies have been primarily carried out in primates and rabbits (Maruoka et al. 2006; Swindle-Reilly et al. 2009). These studies were designed to help in re-attaching the retina after detachment and further studies are needed to determine if they can restore low oxygen tension at the surface of the lens.

2.4   Secondary Cataract

Secondary cataract, also known as PCO, is one of the most common complications following cataract surgery. In PCO, LECs, which remain within the capsule after cataract surgery, are triggered to proliferate and transdifferentiate (West-Mays and Sheardown 2010; Wormstone et al. 2009). Cells derived from the anterior lens epithelium, referred to as "A cells," are those thought to transdifferentiate into spindle-shaped myofibroblasts through a process known as epithelial-to-mesenchymal transformation (EMT) (West-Mays and Sheardown 2010; Wormstone et al. 2009). These cells migrate to the posterior lens capsule where they deposit matrix and cause wrinkling of the capsule, resulting in disruption of vision. PCO was initially diagnosed following the beginning of extracapsular cataract extraction (ECCE) and remained fairly common during these early days (late 1970s and early 1980s) with incidence in up to 50% of patients (Bullimore and Bailey 1993; Kappelhof and Vrensen 1992). However, advances in intraocular lens (IOL) design and surgical technique over the past 20 years have resulted in a dramatic reduction in reported PCO rates, to occurrence in 14–18% of patients (Schaumberg et al. 1998; Wormstone 2002; West-Mays and Sheardown 2010; Wormstone et al. 2009). Yet, PCO remains a major medical problem with profound consequences for the patient's well-being and is a significant financial burden due to the costs of follow-up treatment.

Multiple experimental animal models of PCO have been created, including in vivo, in vitro, and ex vivo approaches (West-Mays and Sheardown 2010; Wormstone et al. 2009) (Table 2.1). The in vitro studies generally involve seeding primary LECs obtained from animal lenses or cell lines created from animal lenses, onto structures such as plexiglass, plastic, or bovine lens capsules with or without IOLs to determine their effects on proliferation and migration. This cell culture approach is the simplest method that has been used to identify which factors can stimulate or inhibit proliferation, migration, differentiation, and transdifferentiation of LEC into a myofibroblast cell phenotype. Cell cultures are often grown on matrices that differ from their native basement membrane (capsule) and this can alter the growth rate and the molecular characteristics of the cells. Thus, to circumvent the phenotypic variation attributed to different cell culture growth surfaces, lens epithelium explant cultures were developed and have been used for some time to investigate lens cell behavior in association with PCO (West-Mays et al. 2010). In particular, rat lens explant studies revealed that TGF is a key growth factor in initiating changes in LECs that are characteristic features of the EMT observed during PCO including a rapid elongation of cells, the aberrant accumulation of ECM, alpha smooth muscle actin (SMA) reactivity, lens capsule wrinkling, and cell death by apoptosis (Hales et al. 1994; Liu et al. 1994; Wallentin et al. 1998). These findings were first highlighted using rat lens explants and this effect of TGF has since been confirmed and extended using a variety of additional species including the mouse, rabbit, bovine, and canine (Mansfield et al. 2004).

Another experimental animal system that has been established, which replicates the in vivo PCO situation closely, is the capsular bag model. These involve performing sham cataract operations on lenses, leaving the bag and remaining

cells behind. Although these were first developed using human donor lenses they have since been applied to bovine, canine, and more recently, rabbit and chick lenses (Hales et al. 1994; Liu et al. 1994; Wormstone et al. 1997; Saxby et al. 1998; Davidson et al. 2000; Walker et al. 2007; Chandler et al. 2012; Pot et al. 2009). The capsular bag model, like the lens explant model, is used to monitor LEC migration as it occurs during PCO from the anterior equatorial margin onto the posterior capsule .

With respect to in vivo animal models for studying PCO, the rabbit is the most commonly used species. This is due to the fact that the rabbit eye is relatively large and has been useful in the assessment of new technologies for removal of the lens/cataract such as surgical blades, phacoemulsification systems, IOLs, IOL insertion systems, ocular irrigating solutions, ophthalmic viscosurgical devices (OVDs), and other novel technologies (Gwon 2008). The rabbit has also been used to test potential therapeutics for preventing PCO, such as microsphere delivery of cyclosporine A (Pei et al. 2013). However, unlike the human lens, the rabbit lens has the ability to regenerate (Gwon 2006), and thus the mechanisms underlying PCO in rabbits may be somewhat different than that which occurs in other mammals whose lenses do not easily regenerate.

More recently, a mouse model of ECCE surgery has been developed to study the pathogenesis of PCO (Manthey et al. 2014). While this model does not involve a replacement with an IOL as done in human corrective surgeries and the rabbit model described above, it does offer the ability to examine the molecular changes in the cells left behind on the capsule following surgery. In addition, transgenic and knockout models can be utilized to determine specific proteins involved. For example, recent work using the ECCE mouse model revealed that the Zeb proteins, Smad interacting protein 1 (Sip1), and -crystallin enhancer-binding factor 1 (dEF1) play important, yet distinct roles during PCO (Manthey et al. 2014).

As outlined above, PCO is considered a proliferative, fibrotic disorder that involves the aberrant deposition of matrix and wrinkling of the lens capsule. Another related cataract, fibrotic cataract, is anterior subcapsular cataract (ASC). ASCs are much less common but can be induced in a pathological situation, that is, it can occur after ocular trauma, surgery, or systemic diseases such as atopic dermatitis and retinitis pigmentosa. Unlike PCO, ASCs are primary cataracts that occur when anterior LECs, in situ, are stimulated to transition into myofibroblasts (Font and Brownstein 1974; Novotny and Pau 1984). In vivo models of TGFβ-induced ASCs have also been developed, including a transgenic mouse model in which active TGFβ is ectopically expressed in lens fiber cells under the control of the αA-crystallin promoter (Srinivasan et al. 1998) and a model involving intravitreal injection of adenovirally expressed TGF (Robertson et al. 2007). In these mice, the ASCs formed closely resemble those observed in humans. An additional ex vivo model includes excised rat lenses that when cultured with TGFβ develop distinct ASC plaques within 6 days that closely mimic human ASC.

2.5 Conclusion

Although much important information on the function and cause of cataracts can be obtained from human epidemiological and postoperative studies, the use of animal models has enabled us to obtain a more comprehensive understanding of the mechanisms of cataractogenesis. Rodents, such as the mouse, rat, and guinea pig, will continue to be the most commonly used models for understanding mechanisms and genetics underlying cataracts due to the fact that they are easy to house and manipulate and are cost-effective. In addition, the mouse and, more recently, the zebra fish represent valuable genetic models for investigating cataracts. Larger mammals such as rabbits, dogs, and also primates have also been utilized, but mainly for screening potential preventative cataract therapeutics. While a number of human LEC lines exist for testing the cellular and biochemical mechanisms of lens cell pathology, the underlying causes of cataractogenesis can only truly be investigated using the "whole

lens" condition due to the importance of its three-dimensional structure in maintaining lens transparency. As a result, it will be difficult to entirely replace animal models of cataracts with human cell lines or capsular bags.

Compliance with Ethical Requirements Judith West-Mays, Scott Bowman and Yizhi Liu declare that they have no conflict of interest.

No human or animal studies were performed by the authors for this article.

References

Al-Ghoul KJ, Kirk T, Kuszak AJ, Zoltoski RK, Shiels A, Kuszak JR (2003) Lens structure in MIP-deficient mice. Anat Rec A Discov Mol Cell Evol Biol 273(2):714–730. doi:10.1002/ar.a.10080

Alizadeh A, Clark JI, Seeberger T, Hess J, Blankenship T, Spicer A, FitzGerald PG (2002) Targeted genomic deletion of the lens-specific intermediate filament protein CP49. Invest Ophthalmol Vis Sci 43(12):3722–3727

Allen PJ (2007) Cataract surgery practice and endophthalmitis prevention by Australian and New Zealand ophthalmologists—comment. Clin Exp Ophthalmol 35(4):391; author reply 391. doi:10.1111/j.1442-9071.2007.01497.x

Bantseev V, Oriowo OM, Giblin FJ, Leverenz VR, Trevithick JR, Sivak JG (2004) Effect of hyperbaric oxygen on guinea pig lens optical quality and on the refractive state of the eye. Exp Eye Res 78(5):925–931. doi:10.1016/j.exer.2004.01.002

Beby F, Commeaux C, Bozon M, Denis P, Edery P, Morle L (2007) New phenotype associated with an Arg-116Cys mutation in the CRYAA gene: nuclear cataract, iris coloboma, and microphthalmia. Arch Ophthalmol 125(2):213–216. doi:10.1001/archopht.125.2.213

Beebe DC, Holekamp NM, Shui YB (2010a) Oxidative damage and the prevention of age-related cataracts. Ophthalm Res 44(3):155–165. doi:10.1159/000316481

Beebe DC, Shui Y-B, Holekamp NM (2010b) Biochemical mechanisms of age-related cataract. In: Levin LA, Albert DM (ed) Ocular diseases. Mechanisms and management. Elsevier Inc., Saunders

Berthoud VM, Beyer EC (2009) Oxidative stress, lens gap junctions, and cataracts. Antioxid Redox Signal 11(2):339–353. doi:10.1089/ars.2008.2119

Black RL, Oglesby RB, Von Sallmann L, Bunim JJ (1960) Posterior subcapsular cataracts induced by corticosteroids in patients with rheumatoid arthritis. JAMA 174:166–171

Bloemendal H, Zweers A, Vermorken F, Dunia I, Benedetti EL (1972) The plasma membranes of eye lens fibres. Biochemical and structural characterization. Cell Differ 1(2):91–106

Boyle DL, Takemoto L, Brady JP, Wawrousek EF (2003) Morphological characterization of the Alpha A- and Alpha B-crystallin double knockout mouse lens. BMC Ophthalmol 3:3

Brady JP, Garland D, Duglas-Tabor Y, Robison WG, Jr., Groome A, Wawrousek EF (1997) Targeted disruption of the mouse alpha A-crystallin gene induces cataract and cytoplasmic inclusion bodies containing the small heat shock protein alpha B-crystallin. Proc Natl Acad Sci U S A 94(3):884–889

Brady JP, Garland DL, Green DE, Tamm ER, Giblin FJ, Wawrousek EF (2001) AlphaB-crystallin in lens development and muscle integrity: a gene knockout approach. Invest Ophthalmol Vis Sci 42(12):2924–2934

Brakenhoff RH, Aarts HJ, Schuren F, Lubsen NH, Schoenmakers JG (1992) The second human beta B2-crystallin gene is a pseudogene. Exp Eye Res 54(5):803–806

Bullimore MA, Bailey IL (1993) Considerations in the subjective assessment of cataract. Optom Vis Sci: Official Publication Am Acad Optometry 70(11):880–885

Capetanaki Y, Smith S, Heath JP (1989) Overexpression of the vimentin gene in transgenic mice inhibits normal lens cell differentiation. J Cell Biol 109(4 Pt 1):1653–1664

Cartier M, Breitman ML, Tsui LC (1992) A frameshift mutation in the gamma E-crystallin gene of the Elo mouse. Nat Genetics 2(1):42–45. doi:10.1038/ng0992-42

Chambers C, Russell P (1991) Deletion mutation in an eye lens beta-crystallin. An animal model for inherited cataracts. J Biol Chem 266(11):6742–6746

Chandler HL, Haeussler DJ, Jr., Gemensky-Metzler AJ, Wilkie DA, Lutz EA (2012) Induction of posterior capsule opacification by hyaluronic acid in an ex vivo model. Invest Ophthalmol Vis Sci 53(4):1835–1845. doi:10.1167/iovs.11-8735

Chang B, Hawes NL, Smith RS, Heckenlively JR, Davisson MT, Roderick TH (1996) Chromosomal localization of a new mouse lens opacity gene (lop18). Genomics 36(1):171–173. doi:10.1006/geno.1996.0439

Cherfan GM, Michels RG, de Bustros S, Enger C, Glaser BM (1991) Nuclear sclerotic cataract after vitrectomy for idiopathic epiretinal membranes causing macular pucker. Am J Ophthalmol 111(4):434–438

Colucci-Guyon E, Portier MM, Dunia I, Paulin D, Pournin S, Babinet C (1994) Mice lacking vimentin develop and reproduce without an obvious phenotype. Cell 79(4):679–694

Coulombre AJ, Coulombre JL (1964) Lens development. I. Role of the lens in eye growth. J Exp Zoology 156:39–47

Davidson MG, Wormstone M, Morgan D, Malakof R, Allen J, McGahan MC (2000) Ex vivo canine lens capsular sac explants. Graefes Arch Clin Exp Ophthalmol 238(8):708–714

de Bustros S, Thompson JT, Michels RG, Enger C, Rice TA, Glaser BM (1988) Nuclear sclerosis after vitrectomy for idiopathic epiretinal membranes. Am J Ophthalmol 105(2):160–164

Delcourt C, Cristol JP, Tessier F, Leger CL, Michel F, Papoz L (2000) Risk factors for cortical, nuclear, and posterior subcapsular cataracts: the POLA study. Pathologies Oculaires Liees a l'Age. Am J Epidemiol 151(5):497–504

DeRosa AM, Xia CH, Gong X, White TW (2007) The cataract-inducing S50P mutation in Cx50 dominantly alters the channel gating of wild-type lens connexins. J Cell Sci 120(Pt 23):4107–4116. doi:10.1242/jcs.012237

DeRosa AM, Mese G, Li L, Sellitto C, Brink PR, Gong X, White TW (2009) The cataract causing Cx50-S50P mutant inhibits Cx43 and intercellular communication in the lens epithelium. Exp Cell Res 315(6):1063–1075. doi:10.1016/j.yexcr.2009.01.017

Devi RR, Yao W, Vijayalakshmi P, Sergeev YV, Sundaresan P, Hejtmancik JF (2008) Crystallin gene mutations in Indian families with inherited pediatric cataract. Mol Vis 14:1157–1170

Djabali K, de Nechaud B, Landon F, Portier MM (1997) AlphaB-crystallin interacts with intermediate filaments in response to stress. J Cell Sci 110(Pt 21):2759–2769

Ehling UH, Charles DJ, Favor J, Graw J, Kratochvilova J, Neuhauser-Klaus A, Pretsch W (1985) Induction of gene mutations in mice: the multiple endpoint approach. Mutation Res 150(1/2):393–401

Eshaghian J, Streeten BW (1980) Human posterior subcapsular cataract. An ultrastructural study of the posteriorly migrating cells. Arch Ophthalmol 98(1):134–143

FitzGerald PG (2009) Lens intermediate filaments. Exp Eye Res 88(2):165–172. doi:10.1016/j.exer.2008.11.007

Font R, Brownstein SA (1974) A light and electron microscopic study of anterior subcapsular cataracts. Am J Ophthalmol 78:972–984

Galichanin K, Lofgren S, Bergmanson J, Soderberg P (2010) Evolution of damage in the lens after in vivo close to threshold exposure to UV-B radiation: cytomorphological study of apoptosis. Exp Eye Res 91(3):369–377. doi:10.1016/j.exer.2010.06.009

Garner WH, Garner MH, Spector A (1981) Gamma-crystallin, a major cytoplasmic polypeptide disulfide linked to membrane proteins in human cataract. Biochem Biophys Res Commun 98(2):439–447

Gehlbach P, Hose S, Lei B, Zhang C, Cano M, Arora M, Neal R, Barnstable C, Goldberg MF, Zigler JS, Jr., Sinha D (2006) Developmental abnormalities in the Nuc1 rat retina: a spontaneous mutation that affects neuronal and vascular remodeling and retinal function. Neuroscience 137(2):447–461. doi:10.1016/j.neuroscience.2005.08.084

Giblin FJ (2000) Glutathione: a vital lens antioxidant. J Ocular Pharmacol Therapeut: Assoc Ocular Pharmacol Therapeut 16(2):121–135

Giblin FJ, Reddy VN (1980) Pyridine nucleotides in ocular tissues as determined by the cycling assay. Exp Eye Res 31(5):601–609

Giblin FJ, Padgaonkar VA, Leverenz VR, Lin LR, Lou MF, Unakar NJ, Dang L, Dickerson JE, Jr., Reddy VN (1995) Nuclear light scattering, disulfide formation and membrane damage in lenses of older guinea pigs treated with hyperbaric oxygen. Exp Eye Res 60(3):219–235

Giblin FJ, Leverenz VR, Padgaonkar VA, Unakar NJ, Dang L, Lin LR, Lou MF, Reddy VN, Borchman D, Dillon JP (2002) UVA light in vivo reaches the nucleus of the guinea pig lens and produces deleterious, oxidative effects. Exp Eye Res 75(4):445–458

Giblin FJ, Quiram PA, Leverenz VR, Baker RM, Dang L, Trese MT (2009) Enzyme-induced posterior vitreous detachment in the rat produces increased lens nuclear pO_2 levels. Exp Eye Res 88(2):286–292. doi:10.1016/j.exer.2008.09.003

Giblin FJ, Lin LR, Simpanya MF, Leverenz VR, Fick CE (2012) A Class I UV-blocking (senofilcon A) soft contact lens prevents UVA-induced yellow fluorescence and NADH loss in the rabbit lens nucleus in vivo. Exp Eye Res 102:17–27. doi:10.1016/j.exer.2012.06.007

Glass AS, Dahm R (2004) The zebrafish as a model organism for eye development. Ophthal Res 36(1):4–24. doi:10.1159/000076105

Goishi K, Shimizu A, Najarro G, Watanabe S, Rogers R, Zon LI, Klagsbrun M (2006) AlphaA-crystallin expression prevents gamma-crystallin insolubility and cataract formation in the zebrafish cloche mutant lens. Development 133(13):2585–2593. doi:10.1242/dev.02424

Gong X, Li E, Klier G, Huang Q, Wu Y, Lei H, Kumar NM, Horwitz J, Gilula NB (1997) Disruption of alpha3 connexin gene leads to proteolysis and cataractogenesis in mice. Cell 91(6):833–843

Gong X, Agopian K, Kumar NM, Gilula NB (1999) Genetic factors influence cataract formation in alpha 3 connexin knockout mice. Dev Gen 24(1/2):27–32. doi:10.1002/(SICI)1520-6408(1999)24:1/2<27::AID-DVG4>3.0.CO;2-7

Goodenough DA (1992) The crystalline lens. A system networked by gap junctional intercellular communication. Seminars Cell Biol 3(1):49–58

Gorin MB, Yancey SB, Cline J, Revel JP, Horwitz J (1984) The major intrinsic protein (MIP) of the bovine lens fiber membrane: characterization and structure based on cDNA cloning. Cell 39(1):49–59

Gosselin ME, Kapustij CJ, Venkateswaran UD, Leverenz VR, Giblin FJ (2007) Raman spectroscopic evidence for nuclear disulfide in isolated lenses of hyperbaric oxygen-treated guinea pigs. Exp Eye Res 84(3):493–499. doi:10.1016/j.exer.2006.11.002

Graw J (1997) The crystallins: genes, proteins and diseases. Biol Chem 378(11):1331–1348

Graw J (2009) Genetics of crystallins: cataract and beyond. Exp Eye Res 88(2):173–189. doi:10.1016/j.exer.2008.10.011

Graw J, Jung M, Loster J, Klopp N, Soewarto D, Fella C, Fuchs H, Reis A, Wolf E, Balling R, Hrabe de Angelis M (1999) Mutation in the betaA3/A1-crystallin encoding gene Cryba1 causes a dominant cataract in the mouse. Genomics 62(1):67–73. doi:10.1006/geno.1999.5974

Graw J, Klopp N, Illig T, Preising MN, Lorenz B (2006) Congenital cataract and macular hypoplasia in humans associated with a de novo mutation in CRYAA and compound heterozygous mutations in P. Graefes Arch Clin Exp Ophthalmol 244(8):912–919. doi:10.1007/s00417-005-0234-x

Graw J, Loster J, Soewarto D, Fuchs H, Meyer B, Reis A, Wolf E, Balling R, Hrabe de Angelis M (2001) Characterization of a new, dominant V124E mutation in the mouse alphaA-crystallin-encoding gene. Invest Ophthalmol Vis Sci 42(12):2909–2915

Gu F, Luo W, Li X, Wang Z, Lu S, Zhang M, Zhao B, Zhu S, Feng S, Yan YB, Huang S, Ma X (2008) A novel mutation in AlphaA-crystallin (CRYAA) caused autosomal dominant congenital cataract in a large Chinese family. Human Mutat 29(5):769. doi:10.1002/humu.20724

Gupta V, Wagner BJ (2009) Search for a functional glucocorticoid receptor in the mammalian lens. Exp Eye Res 88(2):248–256. doi:10.1016/j.exer.2008.04.003

Gwon A (2006) Lens regeneration in mammals: a review. Surv Ophthalmol 51(1):51–62. doi:10.1016/j.survophthal.2005.11.005

Gwon A (2008) The rabbit in cataract/IOL surgery. In: Tsonis P (ed) Animal models in eye research. Elsevier, New York

Halder N, Joshi S, Gupta SK (2003) Lens aldose reductase inhibiting potential of some indigenous plants. J Ethnopharmacol 86(1):113–116

Hales AM, Schulz MW, Chamberlain CG, McAvoy JW (1994) TGF-beta 1 induces lens cells to accumulate alpha-smooth muscle actin, a marker for subcapsular cataracts. Curr Eye Res 13(12):885–890

Hansen L, Yao W, Eiberg H, Kjaer KW, Baggesen K, Hejtmancik JF, Rosenberg T (2007) Genetic heterogeneity in microcornea-cataract: five novel mutations in CRYAA, CRYGD, and GJA8. Invest Ophthalmol Vis Sci 48(9):3937–3944. doi:10.1167/iovs.07-0013

Hatters DM, Lindner RA, Carver JA, Howlett GJ (2001) The molecular chaperone, alpha-crystallin, inhibits amyloid formation by apolipoprotein C-II. J Biol Chem 276(36):33755–33761. doi:10.1074/jbc.M105285200

Hegde KR, Henein MG, Varma SD (2003) Establishment of mouse as an animal model for study of diabetic cataracts: biochemical studies. Diabetes Obes Metab 5(2):113–119

Hejtmancik JF (2008) Congenital cataracts and their molecular genetics. Seminars Cell Dev Biol 19(2):134–149. doi:10.1016/j.semcdb.2007.10.003

Horwitz J (2003) Alpha-crystallin. Exp Eye Res 76(2):145–153

Kador PF, Fukui HN, Fukushi S, Jernigan HM, Jr., Kinoshita JH (1980) Philly mouse: a new model of hereditary cataract. Exp Eye Res 30(1):59–68

Kador PF, Sun G, Rait VK, Rodriguez L, Ma Y, Sugiyama K (2001) Intrinsic inhibition of aldose reductase. J Ocul Pharmacol Ther: The Official J Assoc Ocular Pharmacol Therapeutics 17(4):373–381. doi:10.1089/108076801753162780

Kador PF, Betts D, Wyman M, Blessing K, Randazzo J (2006) Effects of topical administration of an aldose reductase inhibitor on cataract formation in dogs fed a diet high in galactose. Am J Veterinary Res 67(10):1783–1787. doi:10.2460/ajvr.67.10.1783

Kappelhof JP, Vrensen GF (1992) The pathology of after-cataract. A minireview. Acta Ophthalmol Suppl (205):13–24

Khan AO, Aldahmesh MA, Meyer B (2007) Recessive congenital total cataract with microcornea and heterozygote carrier signs caused by a novel missense CRYAA mutation (R54C). Am J Ophthalmol 144(6):949–952. doi:10.1016/j.ajo.2007.08.005

Kronschlager M, Lofgren S, Yu Z, Talebizadeh N, Varma SD, Soderberg P (2013) Caffeine eye drops protect against UV-B cataract. Exp Eye Res 113:26–31. doi:10.1016/j.exer.2013.04.015

Lassen N, Bateman JB, Estey T, Kuszak JR, Nees DW, Piatigorsky J, Duester G, Day BJ, Huang J, Hines LM, Vasiliou V (2007) Multiple and additive functions of ALDH3A1 and ALDH1A1: cataract phenotype and ocular oxidative damage in Aldh3a1(−/−)/Aldh1a1(−/−) knock-out mice. J Biol Chem 282(35):25668–25676. doi:10.1074/jbc.M702076200

Lee AY, Chung SK, Chung SS (1995) Demonstration that polyol accumulation is responsible for diabetic cataract by the use of transgenic mice expressing the aldose reductase gene in the lens. Proc Natl Acad Sci U S A 92(7):2780–2784

Lee KW, Meyer N, Ortwerth BJ (1999) Chromatographic comparison of the UVA sensitizers present in brunescent cataracts and in calf lens proteins ascorbylated in vitro. Exp Eye Res 69(4):375–384. doi:10.1006/exer.1999.0709

Li XQ, Cai HC, Zhou SY, Yang JH, Xi YB, Gao XB, Zhao WJ, Li P, Zhao GY, Tong Y, Bao FC, Ma Y, Wang S, Yan YB, Lu CL, Ma X (2012) A novel mutation impairing the tertiary structure and stability of gammaC-crystallin (CRYGC) leads to cataract formation in humans and zebrafish lens. Hum Mutat 33(2):391–401. doi:10.1002/humu.21648

Li Q, Yan H, Ding TB, Han J, Shui YB, Beebe DC (2013) Oxidative responses induced by pharmacologic vitreolysis and/or long-term hyperoxia treatment in rat lenses. Curr Eye Res 38(6):639–648. doi:10.3109/02713683.2012.760741

Litt M, Kramer P, LaMorticella DM, Murphey W, Lovrien EW, Weleber RG (1998) Autosomal dominant congenital cataract associated with a missense mutation in the human alpha crystallin gene CRYAA. Hum Mol Genet 7(3):471–474

Liu J, Hales AM, Chamberlain CG, McAvoy JW (1994) Induction of cataract-like changes in rat lens epithelial explants by transforming growth factor beta. Invest Ophthalmol Vis Sci 35(2):388–401

Lou MF (2003) Redox regulation in the lens. Prog Retin Eye Res 22(5):657–682

Lyu J, Kim JA, Chung SK, Kim KS, Joo CK (2003) Alteration of cadherin in dexamethasone-induced cataract organ-cultured rat lens. Invest Ophthalmol Vis Sci 44(5):2034–2040

Mackay DS, Andley UP, Shiels A (2003) Cell death triggered by a novel mutation in the alphaA-crystallin gene underlies autosomal dominant cataract linked to chromosome 21q. Eur J Hum Genet: EJHG 11(10):784–793. doi:10.1038/sj.ejhg.5201046

Malicki J (2000) Genetic analysis of eye development in zebrafish. Results Probl Cell Differ 31:257–282

Mansfield KJ, Cerra A, Chamberlain CG (2004) FGF-2 counteracts loss of TGFbeta affected cells from rat lens explants: implications for PCO (after cataract). Mol Vis 10:521–532

Manthey AL, Terrell AM, Wang Y, Taube JR, Yallowitz AR, Duncan MK (2014) The Zeb proteins deltaEF1 and Sip1 may have distinct functions in lens cells following cataract surgery. Invest Ophthalmol Vis Sci 55(8):5445–5455. doi:10.1167/iovs.14-14845

Martinez-Wittinghan FJ, Sellitto C, Li L, Gong X, Brink PR, Mathias RT, White TW (2003) Dominant cataracts result from incongruous mixing of wild-type lens connexins. J Cell Biol 161(5):969–978. doi:10.1083/jcb.200303068

Maruoka S, Matsuura T, Kawasaki K, Okamoto M, Yoshiaki H, Kodama M, Sugiyama M, Annaka M (2006) Biocompatibility of polyvinylalcohol gel as a vitreous substitute. Curr Eye Res 31(7–8):599–606. doi:10.1080/02713680600813854

Mathias RT, White TW, Gong X (2010) Lens gap junctions in growth, differentiation, and homeostasis. Physiol Rev 90(1):179–206. doi:10.1152/physrev.00034.2009

Matsuura T, Yamagishi S, Kodama Y, Shibata R, Ueda S, Narama I (2005) Otsuka Long-Evans Tokushima fatty (OLETF) rat is not a suitable animal model for the study of angiopathic diabetic retinopathy. Int J Tissue Reactions 27(2):59–62

McCarty CA, Taylor HR (2002) A review of the epidemiologic evidence linking ultraviolet radiation and cataracts. Dev Ophthalmol 35:21–31

Meehan S, Berry Y, Luisi B, Dobson CM, Carver JA, MacPhee CE (2004) Amyloid fibril formation by lens crystallin proteins and its implications for cataract formation. J Biol Chem 279(5):3413–3419. doi:10.1074/jbc.M308203200

Merriam JC, Lofgren S, Michael R, Soderberg P, Dillon J, Zheng L, Ayala M (2000) An action spectrum for UV-B radiation and the rat lens. Invest Ophthalmol Vis Sci 41(9):2642–2647

Meyer LM, Lofgren S, Ho YS, Lou M, Wegener A, Holz F, Soderberg P (2009) Absence of glutaredoxin1 increases lens susceptibility to oxidative stress induced by UVR-B. Exp Eye Res 89(6):833–839. doi:10.1016/j.exer.2009.07.020

Michael R, Vrensen GF, van Marle J, Gan L, Soderberg PG (1998) Apoptosis in the rat lens after in vivo threshold dose ultraviolet irradiation. Invest Ophthalmol Vis Sci 39(13):2681–2687

Moghaddam MS, Kumar PA, Reddy GB, Ghole VS (2005) Effect of Diabecon on sugar-induced lens opacity in organ culture: mechanism of action. J Ethnopharmacol 97(2):397–403. doi:10.1016/j.jep.2004.11.032

Moreau KL, King JA (2012) Protein misfolding and aggregation in cataract disease and prospects for prevention. Trends Mol Med 18(5):273–282. doi:10.1016/j.molmed.2012.03.005

Mori Y, Yokoyama J, Nishimura M, Oka H, Mochio S, Ikeda Y (1992) Development of diabetic complications in a new diabetic strain of rat (WBN/Kob). Pancreas 7(5):569–577

Mörner CT (1894) Untersuchungen der Protein substanzen in den lichtbrechenden medien des Auges. Hoppe Seyler Z Physiol Chem 18(61)

Mou L, Xu JY, Li W, Lei X, Wu Y, Xu G, Kong X, Xu GT (2010) Identification of vimentin as a novel target of HSF4 in lens development and cataract by proteomic analysis. Invest Ophthalmol Vis Sci 51(1):396–404. doi:10.1167/iovs.09-3772

Nakamura M, Russell P, Carper DA, Inana G, Kinoshita JH (1988) Alteration of a developmentally regulated, heat-stable polypeptide in the lens of the Philly mouse. Implications for cataract formation. J Biol Chem 263(35):19218–19221

Nicholl ID, Quinlan RA (1994) Chaperone activity of alpha-crystallins modulates intermediate filament assembly. EMBO J 13(4):945–953

Novotny GE, Pau H (1984) Myofibroblast-like cells in human anterior capsular cataract. Virchows Archiv A. Pathological Anatomy Histopathol 404(4):393–401

Obrosova IG, Chung SS, Kador PF (2010) Diabetic cataracts: mechanisms and management. Diabetes/Metabolism Res Rev 26(3):172–180. doi:10.1002/dmrr.1075

Okamura T, Miyoshi I, Takahashi K, Mototani Y, Ishigaki S, Kon Y, Kasai N (2003) Bilateral congenital cataracts result from a gain-of-function mutation in the gene for aquaporin-0 in mice. Genomics 81(4):361–368

Oriowo OM, Cullen AP, Chou BR, Sivak JG (2001) Action spectrum and recovery for in vitro UV-induced cataract using whole lenses. Invest Ophthalmol Vis Sci 42(11):2596–2602

Ostenson CG, Efendic S (2007) Islet gene expression and function in type 2 diabetes; studies in the Goto-Kakizaki rat and humans. Diabetes, Obes Metab 9(Suppl 2):180–186. doi:10.1111/j.1463-1326.2007.00787.x

Palmquist BM, Philipson B, Barr PO (1984) Nuclear cataract and myopia during hyperbaric oxygen therapy. Br J Ophthalmol 68 (2):113–117

Parker NR, Jamie JF, Davies MJ, Truscott RJ (2004) Protein-bound kynurenine is a photosensitizer of oxidative damage. Free Radical Biology Med 37 (9):1479–1489. doi:10.1016/j.freeradbiomed.2004.07.015

Parthasarathy G, Ma B, Zhang C, Gongora C, Samuel Zigler J, Jr., Duncan MK, Sinha D (2011) Expression of betaA3/A1-crystallin in the developing and adult rat eye. J Mol Histol 42(1):59–69. doi:10.1007/s10735-010-9307-1

Pei C, Xu Y, Jiang JX, Cui LJ, Li L, Qin L (2013) Application of sustained delivery microsphere of cyclosporine A for preventing posterior capsular opacification in rabbits. IntJ Ophthalmol 6(1):1–7. doi:10.3980/j.issn.2222-3959.2013.01.01

Pitts DG (1978) Glenn A. Fry Award Lecture–1977. The ocular effects of ultraviolet radiation. Am J Optomet Physiol Optics 55(1):19–35

Pollreisz A, Schmidt-Erfurth U (2010) Diabetic cataractpathogenesis, epidemiology and treatment. J Ophthalmol 2010:608751. doi:10.1155/2010/608751

Posner M (2003) A comparative view of alpha crystallins: the contribution of comparative studies to understanding function. Integr Comp Biol 43(4):481–491. doi:10.1093/icb/43.4.481

Pot SA, Chandler HL, Colitz CM, Bentley E, Dubielzig RR, Mosley TS, Reid TW, Murphy CJ (2009) Selenium functionalized intraocular lenses inhibit posterior capsule opacification in an ex vivo canine lens capsular bag assay. Exp Eye Res 89(5):728–734. doi:10.1016/j.exer.2009.06.016

Pras E, Frydman M, Levy-Nissenbaum E, Bakhan T, Raz J, Assia EI, Goldman B, Pras E (2000) A nonsense mutation (W9X) in CRYAA causes autosomal recessive cataract in an inbred Jewish Persian family. Invest Ophthalmol Vis Sci 41(11):3511–3515

Quiram PA, Leverenz VR, Baker RM, Dang L, Giblin FJ, Trese MT (2007) Microplasmin-induced posterior vitreous detachment affects vitreous oxygen levels. Retina 27(8):1090–1096. doi:10.1097/IAE.0b013e3180654229

Ramaekers FC, Osborn M, Schimid E, Weber K, Bloemendal H, Franke WW (1980) Identification of the cytoskeletal proteins in lens-forming cells, a special epitheloid cell type. Exp Cell Res 127(2):309–327

Rao PV, Zigler JS, Jr. (1990) Extremely high levels of NADPH in guinea pig lens: correlation with zeta-crystallin concentration. Biochem Biophys Res Commun 167(3):1221–1228

Reddy MA, Francis PJ, Berry V, Bhattacharya SS, Moore AT (2004) Molecular genetic basis of inherited cataract and associated phenotypes. Surv Ophthalmol 49(3):300–315. doi:10.1016/j.survophthal.2004.02.013

Richter L, Flodman P, Barria von-Bischhoffshausen F, Burch D, Brown S, Nguyen L, Turner J, Spence MA, Bateman JB (2008) Clinical variability of autosomal dominant cataract, microcornea and corneal opacity and novel mutation in the alpha A crystallin gene (CRYAA). Am J Med Gen Part A 146A (7):833–842. doi:10.1002/ajmg.a.32236

Robertson JV, Nathu Z, Najjar A, Dwivedi D, Gauldie J, West-Mays JA (2007) Adenoviral gene transfer of bioactive TGFbeta1 to the rodent eye as a novel model for anterior subcapsular cataract. Mol Vis 13:457–469

Rong P, Wang X, Niesman I, Wu Y, Benedetti LE, Dunia I, Levy E, Gong X (2002) Disruption of Gja8 (alpha8 connexin) in mice leads to microphthalmia associated with retardation of lens growth and lens fiber maturation. Development 129(1):167–174

Rubinos C, Villone K, Mhaske PV, White TW, Srinivas M (2014) Functional effects of Cx50 mutations associated with congenital cataracts. Am J Physiol Cell Physiol 306(3):C212–220. doi:10.1152/ajpcell.00098.2013

Sandilands A, Hutcheson AM, Long HA, Prescott AR, Vrensen G, Loster J, Klopp N, Lutz RB, Graw J, Masaki S, Dobson CM, MacPhee CE, Quinlan RA (2002) Altered aggregation properties of mutant gamma-crystallins cause inherited cataract. EMBO J 21(22):6005–6014

Sandilands A, Prescott AR, Wegener A, Zoltoski RK, Hutcheson AM, Masaki S, Kuszak JR, Quinlan RA (2003) Knockout of the intermediate filament protein CP49 destabilises the lens fibre cell cytoskeleton and decreases lens optical quality, but does not induce cataract. Exp Eye Res 76(3):385–391

Santhiya ST, Soker T, Klopp N, Illig T, Prakash MV, Selvaraj B, Gopinath PM, Graw J (2006) Identification of a novel, putative cataract-causing allele in CRYAA (G98R) in an Indian family. Mol Vis 12:768–773

Sasase T, Ohta T, Masuyama T, Yokoi N, Kakehashi A, Shinohara M (2013) The spontaneously diabetic torii rat: an animal model of nonobese type 2 diabetes with severe diabetic complications. J Diabetes Res 2013:976209. doi:10.1155/2013/976209

Saxby L, Rosen E, Boulton M (1998) Lens epithelial cell proliferation, migration, and metaplasia following capsulorhexis. Br J Ophthalmol 82(8):945–952

Schaumberg DA, Dana MR, Christen WG, Glynn RJ (1998) A systematic overview of the incidence of posterior capsule opacification. Ophthalmology 105(7):1213–1221. doi:10.1016/S0161-6420(98)97023-3

Schmidt RE, Dorsey DA, Beaudet LN, Peterson RG (2003) Analysis of the Zucker Diabetic Fatty (ZDF) type 2 diabetic rat model suggests a neurotrophic role for insulin/IGF-I in diabetic autonomic neuropathy. Am J Pathol 163(1):21–28. doi:10.1016/S0002-9440(10)63626-7

Shiels A, Bassnett S (1996) Mutations in the founder of the MIP gene family underlie cataract development in the mouse. Nat Genet 12(2):212–215. doi:10.1038/ng0296-212

Shiels A, Hejtmancik JF (2007) Genetic origins of cataract. Arch Ophthalmol 125(2):165–173. doi:10.1001/archopht.125.2.165

Shiels A, Mackay D, Ionides A, Berry V, Moore A, Bhattacharya S (1998) A missense mutation in the human connexin50 gene (GJA8) underlies autosomal dominant "zonular pulverulent" cataract, on chromosome 1q. Am J Hum Genet 62(3):526–532. doi:10.1086/301762

Shiels A, Bassnett S, Varadaraj K, Mathias R, Al-Ghoul K, Kuszak J, Donoviel D, Lilleberg S, Friedrich G, Zambrowicz B (2001) Optical dysfunction of the crystalline lens in aquaporin-0-deficient mice. Physiol Genom 7(2):179–186. doi:10.1152/physiolgenomics.00078.2001

Sidjanin DJ, Parker-Wilson DM, Neuhauser-Klaus A, Pretsch W, Favor J, Deen PM, Ohtaka-Maruyama C, Lu Y, Bragin A, Skach WR, Chepelinsky AB, Grimes PA, Stambolian DE (2001) A 76-bp deletion in the Mip gene causes autosomal dominant cataract in Hfi mice. Genomics 74(3):313–319. doi:10.1006/geno.2001.6509

Simpanya MF, Ansari RR, Leverenz V, Giblin FJ (2008) Measurement of lens protein aggregation in vivo using dynamic light scattering in a guinea pig/UVA model for nuclear cataract. Photochem Photobiol 84(6):1589–1595. doi:10.1111/j.1751-1097.2008.00390.x

Sinha D, Hose S, Zhang C, Neal R, Ghosh M, O'Brien TP, Sundin O, Goldberg MF, Robison WG, Jr., Russell P, Lo WK, Samuel Zigler J, Jr. (2005) A spontaneous mutation affects programmed cell death during development of the rat eye. Exp Eye Res 80(3):323–335. doi:10.1016/j.exer.2004.09.014

Sinha D, Klise A, Sergeev Y, Hose S, Bhutto IA, Hackler L, Jr., Malpic-Llanos T, Samtani S, Grebe R, Goldberg MF, Hejtmancik JF, Nath A, Zack DJ, Fariss RN, McLeod DS, Sundin O, Broman KW, Lutty GA, Zigler JS, Jr. (2008) betaA3/A1-crystallin in astroglial cells regulates retinal vascular remodeling during development. Molecular Cell Neurosci 37(1):85–95. doi:10.1016/j.mcn.2007.08.016

Soderberg PG (1990) Experimental cataract induced by ultraviolet radiation. Acta Ophthalmol Suppl (196):1–75

Soderberg PG, Lofgren S, Ayala M, Dong X, Kakar M, Mody V (2002) Toxicity of ultraviolet radiation exposure to the lens expressed by maximum tolerable dose. Dev Ophthalmol 35:70–75

Spector A, Huang RR, Wang GM (1985) The effect of H2O2 on lens epithelial cell glutathione. Curr Eye Res 4(12):1289–1295

Srinivasan Y, Lovicu FJ, Overbeek PA (1998) Lens-specific expression of transforming growth factor beta1 in transgenic mice causes anterior subcapsular cataracts. J Clin Invest 101(3):625–634. doi:10.1172/JCI1360

Stanga PE, Boyd SR, Hamilton AM (1999) Ocular manifestations of diabetes mellitus. Curr Opin Ophthalmol 10(6):483–489

Swindle-Reilly KE, Shah M, Hamilton PD, Eskin TA, Kaushal S, Ravi N (2009) Rabbit study of an in situ forming hydrogel vitreous substitute. Invest Ophthalmol Vis Sci 50(10):4840–4846. doi:10.1167/iovs.08-2891

Talbot WS, Hopkins N (2000) Zebrafish mutations and functional analysis of the vertebrate genome. Genes Dev 14(7):755–762

Taylor HR (1980) The environment and the lens. Br J Ophthalmol 64(5):303–310

Thisse C, Zon LI (2002) Organogenesis–heart and blood formation from the zebrafish point of view. Science 295(5554):457–462. doi:10.1126/science.1063654

Thompson JT, Glaser BM, Sjaarda RN, Murphy RP (1995) Progression of nuclear sclerosis and long-term visual results of vitrectomy with transforming growth factor beta-2 for macular holes. Am J Ophthalmol 119(1):48–54

Ueda Y, Duncan MK, David LL (2002) Lens proteomics: the accumulation of crystallin modifications in the mouse lens with age. Invest Ophthalmol Vis Sci 43(1):205–215

Vanita V, Singh JR, Hejtmancik JF, Nuernberg P, Hennies HC, Singh D, Sperling K (2006) A novel fan-shaped cataract-microcornea syndrome caused by a mutation of CRYAA in an Indian family. Mol Vis 12:518–522

Varma SD, Kinoshita JH (1974) The absence of cataracts in mice with congenital hyperglycemia. Exp Eye Res 19(6):577–582

Varma SD, Hegde KR, Kovtun S (2008) UV-B-induced damage to the lens in vitro: prevention by caffeine. J Ocul Pharmacol Ther: the official J Assoc Ocular Pharmacol Therapeutics 24(5):439–444. doi:10.1089/jop.2008.0035

Walker JL, Wolff IM, Zhang L, Menko AS (2007) Activation of SRC kinases signals induction of posterior capsule opacification. Invest Ophthalmol Vis Sci 48(5):2214–2223. doi:10.1167/iovs.06-1059

Wallentin N, Wickstrom K, Lundberg C (1998) Effect of cataract surgery on aqueous TGF-beta and lens epithelial cell proliferation. Invest Ophthalmol Vis Sci 39(8):1410–1418

Wang J, Lofgren S, Dong X, Galichanin K, Soderberg PG (2010) Evolution of light scattering and redox balance in the rat lens after in vivo exposure to close-to-threshold dose ultraviolet radiation. Acta Ophthalmologica 88(7):779–785. doi:10.1111/j.1755-3768.2009.01826.x

Wang C, Dawes LJ, Liu Y, Wen L, Lovicu FJ, McAvoy JW (2013) Dexamethasone influences FGF-induced responses in lens epithelial explants and promotes the posterior capsule coverage that is a feature of glucocorticoid-induced cataract. Exp Eye Res 111:79–87. doi:10.1016/j.exer.2013.03.006

Watanabe H, Kosano H, Nishigori H (2000) Steroid-induced short term diabetes in chick embryo: reversible effects of insulin on metabolic changes and cataract formation. Invest Ophthalmol Vis Sci 41(7):1846–1852

Watanabe K, Wada K, Ohashi T, Okubo S, Takekuma K, Hashizume R, Hayashi J, Serikawa T, Kuramoto T, Kikkawa Y (2012) A 5-bp insertion in Mip causes recessive congenital cataract in KFRS4/Kyo rats. PloS ONE 7(11):e50737. doi:10.1371/journal.pone.0050737

West-Mays JA, Sheardown H (2010) Posterior capsule opacification. In: Levin LA, Albert DM (ed) Ocular diseases. Mechanisms and management. Elsevier Inc., Saunders

West-Mays JA, Pino G, Lovicu FJ (2010) Development and use of the lens epithelial explant system to study lens differentiation and cataractogenesis. Prog Retin Eye Res 29(2):135–143. doi:10.1016/j.preteyeres.2009.12.001

White TW (2002) Unique and redundant connexin contributions to lens development. Science 295(5553):319–320. doi:10.1126/science.1067582

White TW, Goodenough DA, Paul DL (1998) Targeted ablation of connexin50 in mice results in microphthalmia and zonular pulverulent cataracts. J Cell Biol 143(3):815–825

WHO (1998) Ageing: a public health challenge. http://www.who.int/mediacentre/factsheets/fs135/en/.

WHO (2000) Blindness: Vision 2020—The Global Initiative for the Elimination of Avoidable Blindness. http://www.who.int/mediacentre/factsheets/fs213/en/.

WHO (2012) Visual impairment and blindness. http://www.who.int/mediacentre/factsheets/fs282/en/index.html.

Wood AM, Truscott RJ (1994) Ultraviolet filter compounds in human lenses: 3-hydroxykynurenine glucoside formation. Vis Res 34(11):1369–1374

Wormstone IM (2002) Posterior capsule opacification: a cell biological perspective. Exp Eye Res 74(3):337–347. doi:10.1006/exer.2001.1153

Wormstone IM, Liu CS, Rakic JM, Marcantonio JM, Vrensen GF, Duncan G (1997) Human lens epithelial cell proliferation in a protein-free medium. Invest Ophthalmol Vis Sci 38(2):396–404

Wormstone IM, Wang L, Liu CS (2009) Posterior capsule opacification. Exp Eye Res 88(2):257–269. doi:10.1016/j.exer.2008.10.016

Wu JW, Chen ME, Wen WS, Chen WA, Li CT, Chang CK, Lo CH, Liu HS, Wang SS (2014) Comparative analysis of human gammaD-crystallin aggregation under physiological and low pH conditions. PloS ONE 9(11):e112309. doi:10.1371/journal.pone.0112309

Xia CH, Cheung D, DeRosa AM, Chang B, Lo WK, White TW, Gong X (2006a) Knock-in of alpha3 connexin prevents severe cataracts caused by an alpha8 point mutation. J Cell Sci 119(Pt 10):2138–2144. doi:10.1242/jcs.02940

Xia CH, Liu H, Chang B, Cheng C, Cheung D, Wang M, Huang Q, Horwitz J, Gong X (2006b) Arginine 54 and Tyrosine 118 residues of {alpha}A-crystallin are crucial for lens formation and transparency. Invest Ophthalmol Vis Sci 47(7):3004–3010. doi:10.1167/iovs.06-0178

Xia CH, Chang B, Derosa AM, Cheng C, White TW, Gong X (2012) Cataracts and microphthalmia caused by a Gja8 mutation in extracellular loop 2. PLoS ONE 7(12):e52894. doi:10.1371/journal.pone.0052894

Zeiss CJ (2013) Translational models of ocular disease. Veterinary Ophthalmol 16(Suppl 1):15–33. doi:10.1111/vop.12065

Zhang C, Gehlbach P, Gongora C, Cano M, Fariss R, Hose S, Nath A, Green WR, Goldberg MF, Zigler JS, Jr., Sinha D (2005) A potential role for beta- and gamma-crystallins in the vascular remodeling of the eye. Dev Dyn 234(1):36–47. doi:10.1002/dvdy.20494

Zhang Y, Ouyang S, Zhang L, Tang X, Song Z, Liu P (2010) Oxygen-induced changes in mitochondrial DNA and DNA repair enzymes in aging rat lens.

Mechanisms Ageing Development 131(11/12):666–673. doi:10.1016/j.mad.2010.09.003

Zhang J, Yan H, Lofgren S, Tian X, Lou MF (2012a) Ultraviolet radiation-induced cataract in mice: the effect of age and the potential biochemical mechanism. Invest Ophthalmol Vis Sci 53(11):7276–7285. doi:10.1167/iovs.12-10482

Zhang P, Xing K, Randazzo J, Blessing K, Lou MF, Kador PF (2012b) Osmotic stress, not aldose reductase activity, directly induces growth factors and MAPK signaling changes during sugar cataract formation. Exp Eye Res 101:36–43. doi:10.1016/j.exer.2012.05.007

Zigler JS, Jr. (1990) Animal models for the study of maturity-onset and hereditary cataract. Exp Eye Res 50(6):651–657

Zigler JS, Jr., Rao PV (1991) Enzyme/crystallins and extremely high pyridine nucleotide levels in the eye lens. FASEB J 5(2):223–225

Zigman S, Datiles M, Torczynski E (1979) Sunlight and human cataracts. Invest Ophthalmol Vis Sci 18(5):462–467

Commentary

Yizhi Liu

e-mail: yzliu62@yahoo.com
Zhongshan Ophthalmic Center,
Sun Yat-sen University,
Guangzhou, China

This chapter extensively reviewed the most commonly established cataract animal models classified as congenital, age-related, and secondary cataracts.

The strengths and weaknesses of different animal species are illustrated and compared, which provides as an outline for choosing and establishing suitable cataract models. Congenital cataract models are largely associated with mutated genes encoding crystallins, connexins, and cytoskeletal and membrane protein families. These mutated gene models are essential for functional analysis of each protein isoform individually. Age-related cataract is highly associated with aging in human; however, diabetic induction, ultraviolet radiation, corticosteroids, and oxygen are commonly used to trigger lens opacities mimicking human age-related cataracts. In secondary cataract models, TGFβ and several other proteins contribute distinctively to the formation of PCO. Different secondary cataract models are useful for evaluating novel surgical techniques or discovering therapeutic interventions.

Although animal models cannot fully represent the features of human cataracts, they are indispensable tools to explore the mechanism of cataractogenesis. Clinicians and researchers alike will find this chapter to be a guide for selecting proper cataract animal models according to different investigation purposes.

Animal Models of Glaucoma

Thomas V. Johnson and Stanislav I. Tomarev

3.1 Introduction

Glaucoma represents a heterogeneous group of chronically progressive optic neuropathies. It is characterized by progressive retinal ganglion cell (RGC) death and axon loss manifesting as retinal nerve fiber layer thinning and optic nerve head (ONH) cupping, which functionally result in a characteristic pattern of visual field loss that begins peripherally, sparing central fixation until late in the disease course. Glaucoma is a major heath burden worldwide, projected to affect more than 110 million people by the year 2040 (Tham et al. 2014). Animal glaucoma models are critical to our understanding of the pathogenesis of the disease and the development of new therapies to prevent blindness. Similar to all animal models of disease, glaucoma models are most valuable when they are readily accessible, efficient for experimentation, and share important pathophysiological

and treatment-response characteristics with the human condition, thereby facilitating clinical translation and improving patient outcomes (Casson et al. 2012). In this chapter, we review the most commonly used rodent models of glaucoma while evaluating their strengths and weaknesses in simulating various aspects of human glaucoma.

The most important modifiable risk factor for the onset and progression of glaucoma in humans is IOP; however, it is notable that the 5-year risk of developing glaucoma for people with IOP greater than 21 mmHg is less than 10% (Kass et al. 2002), and at least 30–40% of patients with glaucoma have a pretreatment IOP of less than 21 mmHg (Sommer et al. 1991; Dielemans et al. 1994; Mitchell et al. 1996), a figure that increases significantly for persons of East Asian descent (Iwase et al. 2004; Kim et al. 2011). Nonetheless, therapeutic reduction of IOP reduces the risk of glaucomatous conversion in patients with ocular hypertension (Kass et al. 2002) and of progression in patients with glaucoma and IOP less than 21 mmHg (Collaborative Normal-Tension Glaucoma Study Group 1998a, b). As such, IOP appears to play an important role in the pathogenesis of glaucoma, and so the animal models generally regarded as most applicable to human disease tend to be those which achieve RGC degeneration via experimental elevation of IOP. At least with regard to medications aimed at lowering IOP, ocular hypertensive animal models of glaucoma are better able to identify those drugs that will go on to commercial availability (Stewart et al. 2011). Alternative models that trigger

S. I. Tomarev (✉)
Section on Retinal Ganglion Cell Biology, Laboratory of Retinal Cell and Molecular Biology, National Eye Institute, National Institutes of Health, 6 Center Drive, MSC 0608, Building 6, Room 212, Bethesda, MD 20892, USA
e-mail: tomarevs@nei.nih.gov

T. V. Johnson
Wilmer Eye Institute, Johns Hopkins Hospital, 600 N Wolfe Street, Wilmer B-29, Baltimore, MD 21287, USA

© Springer International Publishing Switzerland 2016
C.-C. Chan (ed.), *Animal Models of Ophthalmic Diseases*, Essentials in Ophthalmology,

RGC death through more direct toxic or trau- matic mechanisms, such as optic nerve crush or transection, intraocular injection of excitotoxic compounds, ischemia/reperfusion, or induced autoimmune models, are useful for understand- ing RGC pathophysiology in general, but may be less directly applicable to clinical glaucoma and will not be discussed here.

3.2 Rodent Glaucoma Models

Mice and rats are attractive model systems for human disease because of their reasonable ap- proximation to human ocular physiology, low cost, ease of maintenance, relatively short life cycle, and amenability to genetic manipulation. Anatomically, rodent and human eyes share many important characteristics including the anatomy of the anterior chamber drainage structures and physiology of aqueous humor dynamics that can be experimentally manipulated to elevate IOP. It should be noted, however, that rodents exhibit a glial rather than fibrous lamina cribrosa. This could be an important physiological difference given that the site of RGC injury in primates and humans is thought to arise from axonal interac- tions with the lamina cribrosa, which has prompt- ed a recent emphasis on clinical evaluation of this tissue (Kim et al. 2013). Another drawback of ro- dent models of ophthalmic disease, in particular, has been the relatively small ocular size making certain procedures and manipulations (i.e., vit- real sampling and injection) technically more difficult than in larger mammals. Nonetheless, rodent models of glaucoma dominate research within the field because of their broad accessi- bility and versatility. The relatively recent advent of rebound tonometry has improved researchers' ability to accurately measure IOP quickly and re- peatedly in awake or anesthetized rodents (John- son et al. 2008; Morrison et al. 2009; Pease et al. 2011), and is another factor that has led to an increased popularity of rodent glaucoma models in recent years.

Rodent glaucoma models can be broadly classified into those with ocular hypertension induced experimentally and those occurring ge- netically. However, these artificial boundaries are now being blurred by newer techniques that induce ocular hypertension by viral gene trans- fer (see below) and by experiments that take ad- vantage of mouse genetics to induce IOP in ani- mals with gene-encoded cell-specific labels and/ or pathogenesis-related mutations of interest. For instance, several mouse strains that express various fluorescent proteins under control of the RGC-specific gene Thy-1 promoter (Feng et al. 2000) are becoming popular for studying RGC morphology and survival in conjunction with IOP-dependent models of experimentally in- duced glaucoma (Raymond et al. 2009; Tsuruga et al. 2012; Williams et al. 2013).

We and others recently reviewed the benefits and limitations of several important rodent mod- els of glaucoma (Pang and Clark 2007; Howell et al. 2008; McKinnon et al. 2009; Johnson and Tomarev 2010). Here, we will briefly revisit the most popular of the ocular hypertensive rodent glaucoma models and discuss more recent devel- opments in the generation of new rodent glau- coma models.

3.3 Episcleral Vein Injection/ Ablation

The earliest rodent models of IOP-dependent glaucoma involved manipulation of the episcleral veins leading to reduced aqueous humor drain- age from the eye. In the mid-1990s, Morrison and colleagues (Morrison et al. 1997) performed in- travenous injection of hypertonic (1.65–1.85M) saline into the episcleral vessels of adult rats, which causes trabecular meshwork (TM) sclero- sis and formation of anterior synechiae, thereby reducing aqueous humor outflow. This leads to an increase in IOP of roughly between 5 and 25 mmHg that lasts for, on average, 1–5 weeks. Cumulative IOP exposure in this model correlates with RGC axon loss and ONH cupping (Chauhan et al. 2002). Limitations of this technique include the technical difficulty of cannulating the episcleral veins with a custom glass microneedle and the relative variability in the extent and duration of IOP elevation, which sometimes requires reinjection. The method has also been adapted to induce ocular hypertension in the smaller eyes of mice; however, IOP elevation was limited to <5 mmHg and required multiple injections in 80 % of eyes (Kipfer-Kauer et al. 2010). In addition, the method has proven useful in demonstrating protection from optic nerve damage by therapeutic IOP-lowering interventions (Morrison et al. 1998). Recent evidence suggests that this method is associated with structural and functional degeneration of outer retinal neurons in addition to RGCs at relatively late time points well after IOP has returned to baseline (16–26 weeks after induction of ocular hypertension), although visual evoked potentials recorded from the visual cortex remained preserved (Georgiou et al. 2014). This may imply a particular utility for this model in studying the late-term effects of advanced glaucoma on non-RGC retinal neurons.

Another method for experimental IOP elevation in rats also first reported in the 1990s is episcleral vein cauterization (Shareef et al. 1995). The IOP response to cauterization is dose-dependent with respect to the number of episcleral veins involved, with one vessel causing no appreciable IOP elevation and four vessels causing IOP elevation to around 60 mmHg with subsequent proptosis and corneal decompensation. Cauterization of two or three vessels, however, results in increase of IOP to 20–30 mmHg that is sustained for at least several weeks and is associated with progressive RGC apoptosis (Garcia-Valenzuela et al. 1995). An analogous method has more recently been applied to mice with comparable results (Ruiz-Ederra and Verkman 2006). Given that ocular blood flow is disrupted in this method, concern exists that retinal vascular congestion may lead to secondary effects beyond that of IOP elevation alone. Still, recent re-evaluations of this method, and variations thereof, have confirmed its relative ease and reproducibility, and its apparent similarity to

human glaucoma with progressive loss of RGC structure and function (Mittag et al. 2000; Bai et al. 2014). The fact that IOP can be maintained at moderate levels (approximately 30 mmHg) for extended periods of time—up to 16 weeks (Urcola et al. 2006)—makes this model especially useful for long-term studies and arguably approximates chronic human glaucoma better than more acute models of IOP elevation.

3.4 Translimbal Laser Photocoagulation

Photocoagulation of the TM and anterior chamber drainage angle structures is a widely employed technique for reducing aqueous humor outflow and increasing IOP in mice and rats. An early study utilized the method of injecting India ink into the anterior chamber to enhance laser uptake into the TM (Ueda et al. 1998), but more recent adaptations of this technique have shown that IOP elevation is attainable without pigmentary pretreatment. Levkovitch-Verbin et al. (Levkovitch-Verbin et al. 2002) compared four different laser photocoagulation approaches and found that circumferential treatment of the anterior chamber angle without treatment of the episcleral veins resulted in IOP elevation initially to the 40–50 mmHg range that slowly dropped back to baseline over about 3 weeks and caused slowly progressive optic nerve damage. Of note, this technique often requires multiple laser treatments to sustain IOP elevation over the course of weeks. An adaptation of this technique involves paracentesis to flatten the anterior chamber and induce iridocorneal contact at the time of laser treatment, which results in sustained IOP elevation for at least 2–3 weeks without the need for retreatment (Biermann et al. 2012). The intensity of laser power that is used for photocoagulation is an important procedural consideration, and one must take into account whether the animal is pigmented or albino, since trabecular and iris pigment increases laser uptake. According to the protocol, pigmented rats generally require 100 mW for 50–300 ms duration, whereas albino rats require 400–600 mW for 500–700 ms duration. An important factor complicating this technique is corneal decompensation that manifests in a subset of animals and may be related to limbal stem cell destruction from laser treatment.

Mice are also amenable to induction of ocular hypertension by translimbal laser photocoagulation. Pigmentary pretreatment with indocyanine green followed by diode photocoagulation leads to IOP elevation for at least 30 days with subsequent loss of RGC structure and function (Grozdanic et al. 2003). Flattening of the anterior chamber via paracentesis appears to be more critical in this species, and leads to longer periods of IOP elevation, with one report suggesting that IOP remains above 20 mmHg for 4 months following photocoagulation (Feng et al. 2013a). Without these adaptations, translimbal laser photocoagulation in the mouse leads to a doubling of IOP that lasts only for 7 days (Fu and Sretavan 2010). Repeated laser photocoagulation 1 week after the first treatment was reported to increase the mean IOP above 20 mmHg in 57 % of treated eyes at 24 weeks. Treated eyes demonstrated a 59 % reduction in the average axon number as well as a significant decrease in the photopic negative ERG response (Yun et al. 2014).

Recent investigations have identified important neuropathological characteristics of this model. RGC loss following laser trabecular meshwork photocoagulation appears to be topographically sectorial (Soto et al. 2011), which is similar to the pattern of degeneration seen in other models, such as the DBA/2J mouse. Translimbal laser photocoagulation of Thy1-YFP animals has demonstrated a topographic pattern for RGC dendritic tree degeneration that precedes changes to cell soma morphology and begins along the retinal vertical axis (Feng et al. 2013b). Whereas retrograde labeling of RGCs via application of tracers such as fluorogold to the superior colliculus or transected optic nerve stump is often employed to identify and quantify cell survival, a proportion of RGCs are only identifiable through alternative immunohistochemical methods (gamma synuclein or phosphorylated neurofilament), likely because of axonal disruption prior to overt cell death (Soto et al. 2011). This proportion appears to be dependent on the degree of RGC degeneration that is triggered and equates to about 10 % with mild degeneration but up to 85 % in cases of severe degeneration (Soto et al. 2011). This may have important implications for the utilization of glaucoma animal models in general, where retrograde RGC labeling is a common method for quantification of neurodegeneration.

3.5 Microbead Injection

The aforementioned glaucoma models involve obstruction to aqueous humor outflow by physical destruction of anatomical drainage structures. Alternative methods aim to obstruct outflow through the introduction of exogenous substances into the outflow pathway while leaving the biological structures intact, which is likely less inflammatory. Weekly injections of chondroitin sulfate into the anterior chamber is capable of raising IOP in rats to more than 20 mmHg and inducing subsequent RGC loss and electrophysiological changes, but is labor intensive (Belforte et al. 2012). An early approach with microbead injection in rats involved anterior chamber injection of 10-μm latex microspheres with or without viscous hydroxypropylmethylcelluose, but this still required multiple injections to attain sustained IOP elevation (Urcola et al. 2006). Direct comparisons to episcleral vein cauterization show that microspheres with or without hydroxypropylmethylcelluose were capable of attaining sustained IOP elevation to moderate levels (30–40 mmHg) for a longer period of time (more than 30 weeks) with a comparable level of RGC loss at comparable time points (Urcola et al. 2006). Subsequent work has refined the technique to define protocols for reproducible IOP elevation in both mice and rats. Calkins and colleagues (Sappington et al. 2010) evaluated beads measuring 5–15 μm and found that 15-μm beads raised IOP by a modest level of 5–10 mmHg in rats for 2 weeks and in mice for 3 weeks following a single injection, with ocular hypertensive periods extended to 8 weeks in rats injected twice. A variant of this technique in mice utilizes a mixture of 1- and 6-μm beads injected into the anterior chamber followed by viscous sodium hyaluronate, which acts to direct the beads into the anterior chamber angle and maintains elevated IOP for approximately 2 weeks following a single injection (Cone et al. 2012). The injection of a mixture of 6- and 10-μm beads was also used for reproducible and persistent IOP elevation in rats (Smedowski et al. 2014). The relative technical simplicity of this model, low cost, and high degree of reproducibility has provided a system in which relatively large experiments can be conducted with efficiency, leading to a recent increase in new data regarding the pathogenesis of glaucoma.

The effects of clinically used topic ocular hypotensive medications have been evaluated using the microbread model, where experiments have demonstrated a strong IOP lowering response to aqueous suppressants but not to medications that increase outflow (Yang et al. 2012). Recent investigations using the microbead model in mice have elucidated important differences in age and strain-specific responses to experimental glaucoma. Microbead injection induces a greater rise in IOP level when performed in 8-month old as compared with 2-month-old C57BL/6 mice, but this does not hold for CD1 mice (Cone et al. 2012). In addition, older C57BL/6 mice appear to be more resistant to RGC soma and axonal degeneration than younger mice of the same strain, even when statistically corrected for differential IOP exposure, and this effect is associated with less ocular axial elongation in response to ocular hypertension (Cone et al. 2010). CD1 mice, on the other hand, tend to lose more axons than other strains, even when normalizing for IOP exposure (Cone et al. 2010), and older CD1 mice are more susceptible to RGC loss at a given IOP level than younger CD1 mice (Steinhart et al. 2014). Individual RGC classes are also differentially susceptible to IOP-associated cell dysfunction and death as shown in this model, with OFF-transient RGCs showing the earliest structural and functional perturbations (Della Santina et al. 2013). Recent studies using this glaucoma model have begun to elucidate an important role for connective tissue biomechanics in the susceptibility of the optic nerve to neurodegeneration in the face

of elevated IOP. Whereas elastin haploinsufficient animals and fibromodulin knockout mice exhibit similar rates of RGC loss as their wild-type (WT) littermates (Steinhart et al. 2014), mice with the Aca23 mutation in collagen 8A2 have eyes that enlarge to sustained ocular hypertension but are resistant to neurodegeneration (Steinhart et al. 2012).

3.6 Other Models of Induced Ocular Hypertension

An alternative method for inducing elevated IOP in rodents includes viral gene transfer. This approach has the added benefit of generally sustained effects on IOP following a single treatment; however, it benefits from increased flexibility over inborn genetic models of glaucoma in that the timing and in some cases extent of IOP elevation can be controlled by the investigator, and contralateral eyes are available to serve as internal controls. Several genes have been shown to increase IOP when transferred to anterior chamber cells. Bone morphogenetic protein-2 was transduced into TM cells by intracameral injection of modified cytomegalovirus (CMV) expressing the gene. This resulted in calcification of the aqueous outflow pathway and elevation of IOP to 20–30 mmHg for at least a month following a single treatment, which was normalized by topical prostaglandin-analog therapy (Buie et al. 2013). Intravitreal injection of adenovirus encoding soluble CD44 reduced outflow facility in mice, which elevated IOP to nearly 30 mmHg for a period of 50 days following treatment (Giovingo et al. 2013). Similarly, viral transduction of anterior chamber outflow tract cells with transforming growth factor-β2 (TGF-β2) results in increased IOP in both mice and rats, although the effect decreased over the course of about 2 weeks in rats and 3 weeks in mice (Shepard et al. 2010). Interestingly, viral transduction of TM cells in mice with herpes thymidine kinase followed by systemic gancyclovir administration has been used to specifically ablate aqueous outflow pathway cells, thereby reducing rather than elevating IOP (Zhang et al.

2014). This model may prove useful for studying methods to regenerate TM, as meshwork cellularity recovered following ablation (Zhang et al. 2014). Corticosteroid treatment has also been explored as a method for inducing ocular hypertensive glaucoma in rodents, stemming from the tendency of steroids to trigger glaucoma in human patients. Whereas subconjunctival injection of triamcinolone reduces outflow facility in mice but does not alter IOP (Kumar et al. 2013), topical application of dexamethasone three times daily has been shown to increase IOP (Zode et al. 2014).The effect, however, was relatively modest with IOP increasing by only 3 mmHg after 2 weeks of treatment and rising to 8 mmHg at 6 weeks. This did lead to detectable RGC death and optic nerve degeneration; however, the labor intensity of the experimental regimen may limit its utility to studying steroid-induced glaucoma specifically. A potentially more versatile model of steroid-induced glaucoma involves dexamethasone delivery by a subcutaneously implanted osmotic mini-pump in mice. This led to an IOP increase by about 2.5 mmHg over 3 weeks, a decreased conventional outflow facility by 52% versus control, and increased fibrillar material in the TM (Overby et al. 2014).

3.7 Genetic Mouse and Rat Models

Several genetic models have been described that mimic different types of glaucoma, including primary open angle glaucoma (POAG) that occurs with or without elevation of IOP (the latter is also sometimes referred to as normal tension glaucoma or NTG), primary angle closure glaucoma (PACG), pseudoexfoliation syndrome often associated with glaucoma, and congenital glaucoma. Genes known to be involved in human glaucoma have very often been targeted for the development of new mouse models of the disease. Interestingly, however, the observed pathological changes detected in mouse eyes after modification of glaucoma-associated genes sometimes differ from those observed in humans with mutations in the corresponding genes, and the boundaries between subclasses of glaucoma produced in mice have sometimes been blurred.

3.8 Primary Open-Angle Glaucoma Models

Several mouse models of glaucoma have been developed using modifications of genes where mutations are known to cause POAG in humans. *MYOCILIN* (MYOC) was the first gene in which mutations were identified that lead to human POAG (Stone et al. 1997). The *MYOC* gene encodes a secreted glycoprotein. Mutant myocilin proteins associated with severe glaucoma in humans are not secreted from cultured cells or from TM into the aqueous humor (Jacobson et al. 2001, Gobeil et al. 2004). Instead, these mutant myocilin proteins accumulate in the endoplasmic reticulum (ER), which leads to deleterious effects and cell death (Joe et al. 2003; Liu and Vollrath 2004). Expression of mutant myocilin also impairs mitochondrial function (He et al. 2009) and makes cells more sensitive to oxidative stress (Joe and Tomarev 2010), which is intimately connected to ER stress. It is not surprising that a number of labs have attempted to express mutant myocilin using different approaches.

Neither null mutations in the *Myoc* gene (Kim et al. 2001) nor overexpression of the WT myocilin protein in eye drainage structures (Gould et al. 2004a) leads to a glaucoma phenotype in mice, although defects in myelination in the optic nerve of *Myoc* null mice have been reported (Kwon et al. 2014).

The mutant form of human myocilin Y437H expression which triggers severe juvenile-onset POAG in humans (Alward et al. 1998) and the analogous mouse myocilin mutant, Y423H, have been used to generate rodent glaucoma models with variable effects noted. Transgenic mice have been produced using bacterial artificial chromosomes containing mutated human or mouse myocilin genes and long flanking sequences (Senatorov et al. 2006; Zhou et al. 2008). The observed pathological changes using this system were quite modest and similar for both human and mouse mutant myocilin. A moderate elevation in IOP (about 2 mmHg at day time and about 3–4 mmHg at night time above WT littermates) was detected only in mice that were older than 12 months but not in younger animals. A 20% reduction in the number of peripheral but not central RGCs and degeneration of axon fibers mainly in the peripheral areas of the optic nerve were observed only in old transgenic mice (Senatorov et al. 2006; Zhou et al. 2008). Interestingly, RGC electrical responsiveness as judged by pattern electroretinogram was reduced before any reduction in the number of RGCs was detected and was more pronounced in aged versus young mice (Chou et al. 2014). In contrast to the above observations, replacement of the endogenous mouse *Myoc* allele with a mutant Y423H allele did not lead to a detectable elevation of IOP or degenerative changes in the retina or optic nerve (Gould et al. 2006). This might be explained by different genetic backgrounds of mice (McDowell et al. 2012) and/or different levels of mutant myocilin produced. Indeed, expression of high levels of mutated human Y437H myocilin in the eye drainage structures and sclera under the control of the CMV promoter in transgenic mice led to elevation of IOP that was detectable as early as 3 months of age (about 3 and 6 mmHg at day and night time, respectively) (Zode et al. 2011). This was accompanied by the loss of RGCs (about 18 % by 3–5 months of age and 30 % by 12–14 months of age) and optic nerve axons. Functional deficits in RGCs were observed as early as 3–5 months of age. Since the expression of mutant (and WT) myocilin was not detected in the retina, pathological changes in the retina of transgenic mice were not attributed to the retinal expression of the transgene. Expression of mutant human myocilin induced ER stress in the drainage structures (Joe and Tomarev 2010; Zode et al. 2011) and induced apoptosis associated with TM loss, while physiological levels of mutant mouse myocilin did not induce ER stress (Gould et al. 2006). Addition of 20 mM of a chemical chaperone, phenylbutyric acid, to the drinking water reduced ER stress and prevented the glaucoma phenotype in transgenic mice expressing mutant human myocilin under the CMV promoter. Phenylbutyric acid decreased intracellular accumulation of myocilin in ER and promoted the secretion of mutated myocilin in the aqueous humor, thus preventing TM cell death (Zode et al. 2011). Phenylbutyric acid has been approved by FDA for the treatment of a number of diseases, and

data obtained with a myocilin-induced mouse model of POAG as well as with a glucocorticoid-induced mouse model of POAG provide the basis for evaluating the use of chemical chaperones in the treatment of myocilin- and glucocorticoid-induced glaucoma in humans (Zode et al. 2011, 2014).

A genetic mouse model based on the synergistic interaction of mutated myocilin and another significant risk factor, oxidative stress, has been recently described. Double-mutant mice were produced bearing human *MYOC* with a Y437H point mutation as well as a heterozygous deletion of the gene for the primary antioxidant enzyme, superoxide dismutase 2 (SOD2). These mice demonstrated more dramatic changes in the retina, optic nerve, and TM compared with mice carrying the *MYOC* mutation alone. Elevation of IOP (about 4 mmHg over WT littermates at day time) was also detected in 10- to 12-month-old double mutant mice, whereas no elevation of IOP was detected in single *MYOC* mutant or SOD2 deficient lines. Based on this interaction, it was suggested that patients with myocilin-induced glaucoma should avoid habits such as smoking that cause free radical production and/or utilize antioxidant supplements to delay the progression of the disease (Joe et al. 2015).

Another transgenic mouse model of POAG involves expression of connective tissue growth factor (CTGF) (Junglas et al. 2012). Although there are no reports describing mutations in CTGF in human glaucoma, it is expressed in abundance in the human TM (Tomarev et al. 2003), and CTGF levels are increased in the aqueous humor of patients with pseudoexfoliation glaucoma (Browne et al. 2011). Moreover, CTGF is modified by the TGF-β2 pathway and activation of TGF-β_2 in the aqueous humor of patients with POAG has been demonstrated (Tripathi et al. 1994; Inatani et al. 2001). Expression of CTGF under the control of the βB1-crystallin promoter led to elevation of IOP in the eyes of 2- to 3-month-old mice from about 17 mmHg in control mice to 21–22 mmHg in transgenic mice. Light microscopy was unable to identify structural differences in the ciliary body, iris, TM, and Schlemm's canal of 1- to 3-month-old transgenic mice when compared with their WT littermates, and the chamber angle was wide open (Junglas et al. 2012). However, a 150- to 200-nm broad area underneath the cell membrane was observed in TM cells of transgenic mice. This area contained bundles of 6- to 7-nm microfilaments corresponding to the diameter of actin filaments. Such bundles were more rarely observed and were considerably thinner in WT littermates (Junglas et al. 2012). Thus, the effects of CTGF on IOP are probably caused by a modification of the TM actin cytoskeleton. Increased IOP in these mice was accompanied by degenerative changes in the optic nerve. A 26 % reduction in the number of optic nerve axons was detected in 3-month-old βB1-CTGF transgenic mice compared with their WT littermates (Junglas et al. 2012). CTGF-overexpressing mice represent a useful model that mimics the essential functional and structural aspects of POAG (Junglas et al. 2012).

Mice lacking the α_1-subunit of the nitric oxide receptor-soluble guanylate cyclase (sGC$\alpha_1^{-/-}$) represent another model of POAG, as they have been shown to undergo thinning of the retinal nerve fiber layer and loss of optic nerve axons without detectable changes in the morphology of the iridocorneal angle (Buys et al. 2013). Nitric oxide-cGMP signaling has been suggested to be involved in the regulation of aqueous humor outflow and IOP (Kotikoski et al. 2003). In the sGC$\alpha_1^{-/-}$ line, aqueous humor outflow rate was reduced in 57-week-old mice, and a statistically significant elevation of IOP was detected starting from 39 weeks of age, while at 73 weeks of age IOP was about 4 mmHg higher than in age-matched WT mice. Retinal arterial dysfunction has also been detected in sGC$\alpha_1^{-/-}$ mice compared with WT mice (Buys et al. 2013). The sGC$\alpha_1^{-/-}$ line might be a useful tool to study involvement of nitric oxide-cGMP signaling in POAG.

As mentioned above, the Aca23 mutation in collagen 8A2 leads to greater globe enlargement in response to ocular hypertension but not to neurodegeneration (Steinhart et al. 2012). In contrast, expression of the mutated α1-subunit of collagen type I (Col1a1) in transgenic mice leads to not only IOP elevation, but also optic nerve dam-

age (Mabuchi et al. 2004). The mean IOP of the transgenic *Col1a1* mutant mice was significantly elevated compared with that of controls at 16–54 weeks by 20–40 %. It has been demonstrated that IOP and outflow facility in these transgenic mice were inversely correlated over the 12- to 56-week study period (Dai et al. 2009). The mean axonal loss between 24 and 54 weeks of age in the transgenic *Col1a1* mice was about 28 % (Mabuchi et al. 2004). *Col1a1* mutant mice may serve as a useful model of POAG and for evaluating the relationship between collagen type 1 metabolism and aqueous outflow (Dai et al. 2009).

Several mouse models of NTG have been developed. Two NTG models are based on the expression of mutated optineurin (Optn) or WD repeat-containing protein 36. Mutations in the *OPTINEURIN (OPTN)* (Rezaie et al. 2002) and *WDR36* (Monemi et al. 2005) genes were originally identified as being associated with POAG. Subsequent publications suggested that *WDT36* may act as a modifier gene for POAG (Hauser et al. 2006a). The *OPTN* gene has also been associated with a number of different diseases including amyotrophic lateral sclerosis and Paget's disease of bone (Albagha et al. 2011). Optn plays a role in the regulation of cell division, membrane trafficking, protein secretion, and host defense against pathogens (Kachaner et al. 2012). The E50K mutation on Optn has been associated with more severe forms of NTG (Aung et al. 2005; Hauser et al. 2006b). Transgenic mouse lines have been created that express the E50K Optn mutant under the control of chicken beta-actin promoter and CMV enhancer (Chi et al. 2010a). These mice did not show statistically significant elevation of IOP even in 16-month-old animals, but did demonstrate loss of RGCs and connecting synapses in the peripheral retina. Expression of the E50K Optn mutant led to the appearance of apoptotic cells and degeneration of the entire retina, leading to approximately a 28 % reduction in transgenic retinal thickness at 16 months of age (Chi et al. 2010a). Transgenic mice also demonstrated reactive gliosis and E50K mutant protein deposits in the outer plexiform layer (Minegishi et al. 2013). The E50K mutation disrupts the interaction between Optn and Rab8 GTPase, a protein involved in the regulation of vesicle transport from the Golgi to the plasma membrane, and it was suggested that disruption of the Optn-Rab8 complex may affect the trafficking in the RGCs as well in the photoreceptors (Chi et al. 2010a). The E50K mutant also demonstrated enhanced interaction with TANK-binding kinase 1 (TBK1) (Minegishi et al. 2013). Mutations in TBK1 or variations in its copy number have been associated with NTG (Fingert et al. 2011; Seo et al. 2013). The cross talk of Optn and TBK1 is likely to play a significant role not only in glaucoma but also in other retinal diseases, and it was suggested that compounds reducing or abolishing the interaction between the E50K mutant and TBK1 are likely to be beneficial in the treatment of NTG patients (Minegishi et al. 2013).

WD repeat-containing protein 36 is involved in the maturation of 18S rRNA. Expression of mutant Wdr36 in transgenic mice under the control of the chicken beta-actin promoter and CMV enhancer led to a 25% reduction in the thickness of the peripheral retina and loss of RGCs of 16-month-old mice compared with WT mice without statistically significant changes in IOP (Chi et al. 2010b). At the same time, mice carrying only a single copy of the *Wdr36* gene did not develop glaucoma and did not show any detectable eye pathology (Gallenberger et al. 2014).

Mice deficient in the glutamate transporters Glast or Eaac1 have also been considered as mouse models of NTG (Harada et al. 2007). These mice demonstrate spontaneous RGC and optic nerve degeneration without elevated IOP. The number of RGCs was reduced by about 50% in 8-month-old *Glast* null mice compared with their WT littermates. A 20% reduction in the number of cells in the inner nuclear layer was also observed in *Glast* null mice of this age. In *Glast*-deficient mice, the glutathione level in Müller glia was decreased and the administration of glutamate receptor blocker or overexpression of dedicator of cytokinesis 3 (Dock3) (Namekata et al. 2013) prevented RGC loss. In EAAC1-deficient mice, RGCs were more vulnerable to oxidative stress. Data obtained with these models suggest that glutamate transporters are necessary both for prevention of excitotoxic retinal damage

and for synthesis of glutathione, a major cellular antioxidant (Harada et al. 2007). Since Dock3 directly binds to the intracellular C-terminus domain of an *N*-methyl-D-aspartate receptor subunit NR2B, it was also proposed that Dock3 overexpression prevents glaucomatous retinal degeneration by suppressing NR2B-mediated glutamate neurotoxicity and oxidative stress (Namekata et al. 2013). In general, these described models of NTG might be used to investigate the mechanisms of RGC death independent of IOP elevation as well as to develop possible therapies that could augment IOP reduction.

It has been proposed that endothelin 1 (EDN1), a potent vasocontractor, may contribute to glaucoma pathophysiology (Prasanna et al. 2011). Overexpression of End1 under the control of the receptor tyrosine kinase tie-1 promoter in transgenic mice led to a progressive loss of RGCs and a decreased thickness of the inner nuclear layer and outer nuclear layer as early as around 10–12 months without elevation of IOP (Mi et al. 2012). Degeneration of optic nerve axons, together with blood vessel changes and retinal gliosis, was also observed in 24-month-old mice (Mi et al. 2012). This model might be useful to study the contribution of endothelial EDN1-related mechanisms to glaucoma.

3.9 Primary Angle-Closure Glaucoma Models

Unlike POAG in which IOP elevation occurs in the presence of structurally unimpeded aqueous outflow pathways, PACG is associated with physical loss of access to the drainage angle structures. The incidence of PACG is higher in Asians than in Europeans and Africans, and PACG causes blindness more frequently than POAG (Quigley and Broman 2006; Tham et al. 2014). Several mouse models of PACG have been developed.

The genes for *VAV2* and *VAV3*, guanine nucleotide exchange factors for Rho guanosine triphosphatases, have been reported to cause POAG in the Japanese (Fujikawa et al. 2010). Although the involvement of *VAV2* and *VAV3* variants in

POAG, NTG, and PACG has not been confirmed by other studies (Rao et al. 2010; Shi et al. 2013), analysis of *Vav2/Vav3* null mice, and to a lesser degree *Vav2* null mice, showed that these mice develop eye pathologies that resemble glaucoma in humans (Fujikawa et al. 2010). Eyes of *Vav2/Vav3* null mice develop buphthalmos and frequently demonstrate angle closure starting between 6 and 12 weeks of age. Six-week-old *Vav2/Vav3* null mice showed elevated IOP (18.2±3.1 vs. 14.0±2.4 mmHg), which increased further by 10 weeks of age (22.5±7.4 vs. 14.6±4.2 mmHg) (Fujikawa et al. 2010). IOP in *Vav2/Vav3* null mice could be reduced by a number of drugs used for IOP reduction in humans. Although quantitative estimates were not presented, *Vav2/Vav3* null mice demonstrated RGC loss and ONH cupping at 10 weeks of age which was more pronounced in aged mice (Fujikawa et al. 2010). The early onset of IOP elevation and high frequency of ocular pathologies makes this model rather attractive for neuroprotective studies as well as for the investigation of molecular mechanisms involved in pathological structural changes.

Another mouse mutant line that mimics PACG, glaucoma relevant mutant 4 or *Grm4,* was identified in an *N*-ethyl-*N*-nitrosourea mutagenesis screen designed to discover new genes that may affect IOP (Nair et al. 2011). *Grm4* mice carry a mutation in the *Prss56* gene encoding serine protease. The identified mutation leads to a truncation of the C-terminal part of the protein, which does not affect protease activity with the synthetic substrate. The homozygous *Prss56* mutants show elevated IOP and the incidence of high IOP (more than 19 vs. 15 mmHg in WT) increased with age (~50% of mice at 3 months and ~90% of mice at 12 months of age) (Nair et al. 2011). The eye angle was narrow and the iris was in close proximity to the TM in mutant mice. In addition, *Grm4* mice had a shorter axial length than control mice (Nair et al. 2011). Mutations on the *PRSS56* gene have also been shown to segregate with disease in human families with posterior microphthalmia (Gal et al. 2011; Nair et al. 2011). *Grm4* mice therefore may serve as a model to study a spectrum of ocular phenotypes, including changes in the size of the eye, hyperopia, and PACG (Nair et al. 2011).

3.10 Pigmentary Dispersion and Exfoliation Glaucoma Models

Certainly, the most widely used and best characterized model of naturally occurring glaucoma is the DBA/2J mouse (Libby et al. 2005a), which bears recessive mutations in the *Gpnmb* and *Tyrp1* genes, encoding tyrosinase-related and glycosylated transmembrane proteins, respectively. These mutations lead to pigment dispersion, iris transillumination, iris atrophy, and anterior synechia (Anderson et al. 2002). The subsequent blockade of aqueous outflow results in ocular hypertension by the age of 9 months, which is accompanied by the typical signs of glaucoma including death of the RGCs, optic nerve atrophy, ONH cupping, and visual deficits (Libby et al. 2005a). Interestingly, DBA/2J mice are more resistant to the RGC death after optic nerve crush than 14 other inbred lines of mice tested. It was suggested that two dominant loci are linked to the resistance phenotype (Li et al. 2007).

Several important observations have been made using the DBA/2J model. It has been shown that degeneration of RGCs does not occur in the uniform manner across the retina. Regions of cell death and survival radiating from the ONH in fan-shaped sectors have been detected (Jakobs et al. 2005; Howell et al. 2007). Early signs of axon damage are localized to an astrocyte-rich glial lamina region of the optic nerve just posterior to the retina (Howell et al. 2007). It has been suggested that axon damage at the ONH might be a primary lesion in this model, similar to the pathophysiology of human glaucoma (Jakobs et al. 2005). The proapoptotic protein BAX was shown to be necessary for RGC death but not for RGC axon degeneration (Libby et al. 2005b). At the same time, deficiency in *Bim*, proapoptotic Bcl-2 family member that induces BAX activation, did not prevent RGC death in DBA/2J eyes with severe optic nerve damage (Harder et al. 2012). It has, therefore, been proposed that BAX may be a candidate human glaucoma susceptibility gene and that characterization of BAX alleles may be used as a predictive value for glaucoma progression (Libby et al. 2005b).

Similar to the results obtained with rat and monkey models (Ahmed et al. 2004; Stasi et al. 2006; Guo et al. 2010), genes involved in glial activation and immune response are activated in DBA/2J retina (Steele et al. 2006; Panagis et al. 2010, 2011). The complement component C1q, the initiating protein of the classical complement cascade, is upregulated in the retina in several animal models of glaucoma and human glaucoma (Stasi et al. 2006). In DBA/2J mice, C1q, and downstream C3 relocalize to adult retinal synapses at an early stage of glaucoma prior to visible signs of neurodegeneration. C1q in the adult glaucomatous retina may mark synapses for elimination at early stages of the disease, suggesting that the complement cascade mediates synapse loss in glaucoma similar to synapse elimination during development (Stevens et al. 2007; Rosen and Stevens, 2010). It has been shown that upregulation of the complement cascade and the endothelin system (Edn2) occurs at early stages of glaucoma before detectable glaucomatous changes (Howell et al. 2011). Disturbing any of these pathways separately through a mutation in C1qa or through inhibition of the endothelin system with bosentan, an inhibitor of endothelin receptor type A, was protective against retinal and optic nerve damage in DBA/2J mice (Howell et al. 2011). Combinatorial targeting of the endothelin system with bosentan and the complement system with a mutant in C1qa was more effective in protecting the optic nerve than separately inhibiting either process (Howell et al. 2014). It is interesting to note that DBA/2J mice are naturally deficient in complement component 5, and backcrossing of a functional C5 gene into DBA/2J leads to a more severe glaucoma at an earlier age than in regular DBA/2J mice (Howell et al. 2013). Novel therapeutic strategies for glaucoma have also been suggested on the basis of the results obtained with DBA/2J mice. It has been demonstrated that high-dose γ-irradiation together with bone marrow transfer protected RGCs in DBA/2J mice (Anderson et al. 2005). It has been later shown by the same group that the entry of proinflammatory monocytes into the DBA/2J optic nerve occurs very early prior to neural damage and a one-time treatment of an individual eye in young mice with X-rays protects a treated eye from glaucoma even without bone marrow transfer. The X-ray treatment modifies activation of endothelial cell signaling and abolishes migration of monocytes into the eye (Howell et al. 2012). At the same time, exposure of rats to head-only X-ray irradiation provided no protection from optic nerve degeneration induced in rats by an episcleral vein injection of hypertonic saline (Johnson et al. 2014). The elucidation of the biological basis of differences in response to radiation in two different models of glaucoma, as well as in other glaucoma models, may be essential for a better understanding of glaucomatous nerve injury in human glaucoma and for improvements of glaucoma treatment (Johnson et al. 2014).

It has been demonstrated that sequence variants in the *LOXL1* gene encoding lysyl oxidase-like protein 1 are associated with pseudoexfoliation glaucoma in humans (Thorleifsson et al. 2007). *Loxl1* null mice show impairment of the blood–aqueous humor barrier in the ocular anterior segment that causes lens abnormalities, but do not demonstrate deposition of macromolecular material or glaucoma (Wiggs et al. 2014). Although *Loxl1* null mice have some features of pseudoexfoliation syndrome suggesting that decreased enzyme activity contributes to predisposition to the disease, these results suggest that complete disease manifestation in mice requires other factors that could be genetic and/or environmental (Wiggs et al. 2014). It was suggested that production of mouse lines using BAC containing high-risk human alleles of *LOXL1* may lead to an improved mouse model of pseudoexfoliation syndrome (John et al. 2014).

Pseudoexfoliation syndrome is partially recapitulated in *Lyst* mutant mice homozygous for the *beige* allele (Trantow et al. 2009). The *Lyst* gene encodes a 430 kD cytosolic protein, and the *beige* mutation leads to a deletion of a single isoleucine from the carboxyl terminus of the Lyst protein within the WD40 domain, potentially disrupting protein–protein interaction (Trantow et al. 2009). Although the *Lyst* mutation does not cause glaucoma in the C57BL/6J genetic background as judged by the absence of IOP increase or optic nerve damage, *Lyst* mice exhibit XFS-like transillumination defects and pronounced pigment dispersion. Data obtained with this model suggest that *LYST-* or *LYST*-interacting genes may contribute to pseudoexfoliation syndrome (Trantow et al. 2009).

3.11 Congenital and Developmental Glaucoma Models

Developmental glaucoma primarily affects tissues of the anterior segment of the eye. They are relatively rare and account for less than 1 % of all human glaucoma cases. Several genes have been implicated in developmental glaucoma and anterior segment dysgenesis in humans, and most of these genes encode transcription factors. They include *FOXC1, FOXC2, FOXE3, PITX2, PITX3, LMX1B, PAX6,* and *MAF* (Cvekl and Tamm 2004; Gould et al. 2004b; Sowden 2007; Liu and Allingham 2011). Knockouts of these genes as a rule are embryonically or neonatally lethal. Conditional knockouts or introduction of mutations in the genes implicated in developmental glaucoma may provide a useful tool to study their function in the development of the eye angle structure and glaucoma.

Mutations in cytochrome P4501B1 (CYP1B1), including truncating mutations, may lead to primary congenital glaucoma (PCG) (Stoilov et al. 1997; Bejjani et al. 1998; Vasiliou and Gonzalez 2008). It has been suggested that diminished or absent metabolism of key endogenous CYP1B1 substrates adversely affects the development of the TM (Vasiliou and Gonzalez 2008). Therefore, *Cyp1b1* null mice have been analyzed for glaucoma phenotypes. It has been shown that *Cyp1b1* null mice have ocular drainage structure abnormalities resembling those reported in human PCG patients (Libby et al. 2003; Zhao et al. 2013; Teixeira et al. 2014). Three-week-old *Cyp1b1*[−/−] mice had a significantly decreased TM collagen than age-matched controls (Zhao et al. 2013; Teixeira et al. 2014). Collagen loss was progressively increased in older animals, with 8-month-old animals presenting severe atrophy of the TM (Teixeira et al. 2014). Elevated oxidative stress was suggested to contribute to the observed pathological changes in the TM of *Cyp1b1* mice. Indeed, administration of the antioxidant *N*-acetylcysteine restored structural abnormality of TM tissues in *Cyp1b1* null mice (Zhao et al. 2013). Detected changes in ocular drainage structures were associated with a modest but statistically significant elevation of IOP, which for mice of 6–12 weeks of age was about 1 mmHg or 10 % higher in *Cyp1b1* null mice than in WT littermates (Zhao et al. 2013).

A rat model of congenital glaucoma derived from the RCS-rdy⁻ strain has also been described (Thanos and Naskar 2004). Mutant rats are affected with either a unilateral or bilateral enlargement of the globes having an IOP that ranged from 25 to 45 mmHg, as compared to control values of 12–16 mmHg. Concomitant with the rise in IOP, the number of labeled RGCs continued to decrease in number with age. The anterior chamber was narrow but the irido-corneal angle in glaucomatous eyes was open. Impairment of aqueous humor outflow via the uveoscleral pathway may contribute to this glaucomatous phenotype (Gatzioufas et al. 2013).

3.12 Future Possible Models

Association studies have identified a number of new POAG loci and disease genes, including *CAV1, CDKN2B, ABCA1, TMCO1, PMM2, AFAP1, GMDS,* and GAS7 (Burdon et al. 2011; Chen et al. 2014; Gharahkhani et al. 2014; Hysi et al. 2014). Modifications of some of these genes have been performed in mice. *Cav1* null mice, for instance, demonstrated a decreased ERG response without histological changes of the retina (Li et al. 2012). However, no studies of IOP, anterior angle, RGCs, and ONH were reported, and further investigation is needed to clarify the potential usage of the *Cav1* null line as a putative POAG model.

A series of recent publications demonstrated that Prox1, a transcription factor containing an unusual homeodomain, plays an important role in the development and function of Schlemm's canal (Kizhatil et al. 2014; Park et al. 2014). It has been suggested that Prox1 may act as a regulator of aqueous humor outflow (Kizhatil et al. 2014). Modification of Prox1 expression in the eye drainage structures or expression of mutated forms of Prox1 may lead to the development of new models of glaucoma.

The recent development of the CRISPR (clustered regularly interspaced short palindromic repeat)–Cas9 (CRISPR-associated nuclease 9) system provides a simple, efficient method to precisely manipulate the genome of virtually any organism (Doudna and Charpentier 2014; Harrison et al. 2014). Efficiency and high specificity of the CRISPR–Cas9 system allows simultaneous modification of several genes (Cong et al. 2013). Using this system, fast manipulation of the mouse and rat genomes with concurrent introduction of mutations in the genes of interest is possible. It is likely that the future will reveal new rat and mouse models of glaucoma developed with the CRISPR–Cas9 system.

3.13   Conclusions

An ideal glaucoma model should be simple to develop, reproducible, and inexpensive, and most importantly as similar to human glaucoma as possible. As discussed, there are a wide variety of mouse and rat models of glaucoma, each with their own relative strengths and weaknesses. Experimentally induced models of ocular hypertension provide the investigator with control over the time course and the extent of IOP elevation but can be labor intensive, and the need for physical manipulation of the eye introduces ocular changes that are not always present in the human condition. Genetic models provide valuable insights into the molecular mechanisms by which glaucoma manifests but require a good deal of effort to generate lines, and the disease manifestations tend to be less pronounced and require longer periods of time to develop than experimentally induced models. Nonetheless, diverse rodent models of glaucoma have provided the scientific community with valuable insights into the pathogenesis of the disease and possible new therapeutic avenues. These insights will no doubt expand as work continues on the ever-expanding repertoire of rodent glaucoma models.

Compliance with Ethical Requirements Thomas V Johnson, Stanislav I Tomarev, and John C. Morrison declare that they have no conflict of interest.

None of the human studies were carried out by the authors for this chapter. All institutional and national guidelines for the care and use of laboratory animals were followed.

References

Ahmed F, Brown KM, Stephan DA, Morrison JC, Johnson EC, Tomarev SI (2004) Microarray analysis of changes in mRNA levels in the rat retina after experimental elevation of intraocular pressure. Invest Ophthalmol Vis Sci 45:1247–1258

Albagha OM, Wani SE, Visconti MR, Alonso N, Goodman K, Brandi ML, Cundy T, Chung PY, Dargie R, Devogelaer JP, Falchetti A, Fraser WD, Gennari L, Gianfrancesco F, Hooper MJ, Van Hul W, Isaia G, Nicholson GC, Nuti R, Papapoulos S, Montes Jdel P, Ratajczak T, Rea SL, Rendina D, Gonzalez-Sarmiento R, Di Stefano M, Ward LC, Walsh JP, Ralston SH, Genetic Determinants of Paget's Disease C (2011) Genome-wide association identifies three new susceptibility loci for Paget's disease of bone. Nat Genet 43:685–689

Alward WL, Fingert JH, Coote MA, Johnson AT, Lerner SF, Junqua D, Durcan FJ, McCartney PJ, Mackey DA, Sheffield VC, Stone EM (1998) Clinical features associated with mutations in the chromosome 1 open-angle glaucoma gene (GLC1A). New Engl J Med 338:1022–1027

Anderson MG, Smith RS, Hawes NL, Zabaleta A, Chang B, Wiggs JL, John SW (2002) Mutations in genes encoding melanosomal proteins cause pigmentary glaucoma in DBA/2J mice. Nat Genet 30:81–85

Anderson MG, Libby RT, Gould DB, Smith RS, John SW (2005) High-dose radiation with bone marrow transfer prevents neurodegeneration in an inherited glaucoma. Proc Natl Acad Sci U S A 102:4566–4571

Aung T, Rezaie T, Okada K, Viswanathan AC, Child AH, Brice G, Bhattacharya SS, Lehmann OJ, Sarfarazi M, Hitchings RA (2005) Clinical features and course of patients with glaucoma with the E50K mutation in the optineurin gene. Invest Ophthalmol Vis Sci 46:2816–2822

Bai Y, Zhu Y, Chen Q, Xu J, Sarunic MV, Saragovi UH, Zhuo Y (2014) Validation of glaucoma-like features in the rat episcleral vein cauterization model. Chin Med J (Engl) 127:359–364

Bejjani BA, Lewis RA, Tomey KF, Anderson KL, Dueker DK, Jabak M, Astle WF, Otterud B, Leppert M, Lupski JR (1998) Mutations in CYP1B1, the gene for cytochrome P4501B1, are the predominant cause of primary congenital glaucoma in Saudi Arabia. Am J Hum Genet 62:325–333

Belforte N, Sande PH, de Zavalia N, Dorfman D, Rosenstein RE (2012) Therapeutic benefit of radial optic neurotomy in a rat model of glaucoma. PloS ONE 7:e34574

Biermann J, van Oterendorp C, Stoykow C, Volz C, Jehle T, Boehringer D, Lagreze WA (2012) Evaluation of intraocular pressure elevation in a modified laser-induced glaucoma rat model. Exp Eye Res 104:7–14.

Browne JG, Ho SL, Kane R, Oliver N, Clark AF, O'Brien CJ, Crean JK (2011) Connective tissue growth factor is increased in pseudoexfoliation glaucoma. Invest Ophthalmol Vis Sci 52:3660–3666

Buie LK, Karim MZ, Smith MH, Borras T (2013) Development of a model of elevated intraocular pressure in rats by gene transfer of bone morphogenetic protein 2. Invest Ophthalmol Vis Sci 54:5441–5455

Burdon KP, Macgregor S, Hewitt AW, Sharma S, Chidlow G, Mills RA, Danoy P, Casson R, Viswanathan AC, Liu JZ, Landers J, Henders AK, Wood J, Souzeau E, Crawford A, Leo P, Wang JJ, Rochtchina E, Nyholt DR, Martin NG, Montgomery GW, Mitchell P, Brown MA, Mackey DA, Craig JE (2011) Genome-wide association study identifies susceptibility loci for open angle glaucoma at TMCO1 and CDKN2B-AS1. Nat Genet 43:574–578

Buys ES, Ko YC, Alt C, Hayton SR, Jones A, Tainsh LT, Ren R, Giani A, Clerte M, Abernathy E, Tainsh RE, Oh DJ, Malhotra R, Arora P, de Waard N, Yu B, Turcotte R, Nathan D, Scherrer-Crosbie M, Loomis SJ, Kang JH, Lin CP, Gong H, Rhee DJ, Brouckaert P, Wiggs JL, Gregory MS, Pasquale LR, Bloch KD, Ksander BR (2013) Soluble guanylate cyclase alpha1-deficient mice: a novel murine model for primary open angle glaucoma. PloS ONE 8:e60156

Casson RJ, Chidlow G, Wood JP, Crowston JG, Goldberg I (2012) Definition of glaucoma: clinical and experimental concepts. Clin Exp Ophthalmol 40:341–349

Chauhan BC, Pan J, Archibald ML, LeVatte TL, Kelly ME, Tremblay F (2002) Effect of intraocular pressure on optic disc topography, electroretinography, and axonal loss in a chronic pressure-induced rat model of optic nerve damage. Invest Ophthalmol Vis Sci 43:2969–2976

Chen Y, Lin Y, Vithana EN, Jia L, Zuo X, Wong TY, Chen LJ, Zhu X, Tam PO, Gong B, Qian S, Li Z, Liu X, Mani B, Luo Q, Guzman C, Leung CK, Li X, Cao W, Yang Q, Tham CC, Cheng Y, Zhang X, Wang N, Aung T, Khor CC, Pang CP, Sun X, Yang Z (2014) Common variants near ABCA1 and in PMM2 are associated with primary open-angle glaucoma. Nat Genet 46:1115–1119

Chi ZL, Akahori M, Obazawa M, Minami M, Noda T, Nakaya N, Tomarev S, Kawase K, Yamamoto T, Noda S, Sasaoka M, Shimazaki A, Takada Y, Iwata T (2010a) Overexpression of optineurin E50K disrupts Rab8 interaction and leads to a progressive retinal degeneration in mice. Hum Mol Genet 19:2606–2615

Chi ZL, Yasumoto F, Sergeev Y, Minami M, Obazawa M, Kimura I, Takada Y, Iwata T (2010b) Mutant WDR36 directly affects axon growth of retinal ganglion cells leading to progressive retinal degeneration in mice. Hum Mol Genet 19:3806–3815

Chou TH, Tomarev S, Porciatti V (2014) Transgenic mice expressing mutated Tyr437His human myocilin develop progressive loss of retinal ganglion cell electrical responsiveness and axonopathy with normal iop. Invest Ophthalmol Vis Sci 55:5602–5609

Collaborative Normal-Tension Glaucoma Study Group (1998a) Comparison of glaucomatous progression between untreated patients with normal-tension glaucoma and patients with therapeutically reduced intraocular pressures. Am J Ophthalmol 126: 487–497

Collaborative Normal-Tension Glaucoma Study Group (1998b) The effectiveness of intraocular pressure reduction in the treatment of normal-tension glaucoma. Am J Ophthalmol 126:498–505

Cone FE, Gelman SE, Son JL, Pease ME, Quigley HA (2010) Differential susceptibility to experimental glaucoma among 3 mouse strains using bead and viscoelastic injection. Exp Eye Res 91:415–424

Cone FE, Steinhart MR, Oglesby EN, Kalesnykas G, Pease ME, Quigley HA (2012) The effects of anesthesia, mouse strain and age on intraocular pressure and an improved murine model of experimental glaucoma. Exp Eye Res 99:27–35

Cong L, Ran FA, Cox D, Lin S, Barretto R, Habib N, Hsu PD, Wu X, Jiang W, Marraffini LA, Zhang F (2013) Multiplex genome engineering using CRISPR/Cas systems. Science 339:819–823

Cvekl A, Tamm ER (2004) Anterior eye development and ocular mesenchyme: new insights from mouse models and human diseases. Bioessays 26:374–386

Dai Y, Lindsey JD, Duong-Polk X, Nguyen D, Hofer A, Weinreb RN (2009) Outflow facility in mice with a targeted type I collagen mutation. Invest Ophthalmol Vis Sci 50:5749–5753

Della Santina L, Inman DM, Lupien CB, Horner PJ, Wong RO (2013) Differential progression of structural and functional alterations in distinct retinal ganglion cell types in a mouse model of glaucoma. J Neurosci 33:17444–17457

Dielemans I, Vingerling JR, Wolfs RC, Hofman A, Grobbee DE, de Jong PT (1994) The prevalence of primary open-angle glaucoma in a population-based study in The Netherlands. The Rotterdam study. Ophthalmology 101:1851–1855

Doudna JA, Charpentier E (2014) Genome editing. The new frontier of genome engineering with CRISPR-Cas9. Science 346:1258096

Feng G, Mellor RH, Bernstein M, Keller-Peck C, Nguyen QT, Wallace M, Nerbonne JM, Lichtman JW, Sanes JR (2000) Imaging neuronal subsets in transgenic mice expressing multiple spectral variants of GFP. Neuron 28:41–51

Feng L, Chen H, Suyeoka G, Liu X (2013a) A laser-induced mouse model of chronic ocular hypertension to characterize visual defects. J Vis Exp J Vis Exp 14:78. doi:10.3791/50440

Feng L, Zhao Y, Yoshida M, Chen H, Yang JF, Kim TS, Cang J, Troy JB, Liu X (2013b) Sustained ocular hypertension induces dendritic degeneration of mouse retinal ganglion cells that depends on cell type and location. Invest Ophthalmol Vis Sci 54:1106–1117

Fingert JH, Robin AL, Stone JL, Roos BR, Davis LK, Scheetz TE, Bennett SR, Wassink TH, Kwon YH, Alward WL, Mullins RF, Sheffield VC, Stone EM (2011) Copy number variations on chromosome 12q14 in patients with normal tension glaucoma. Hum Mol Genet 20:2482–2494

Fu CT, Sretavan D (2010) Laser-induced ocular hypertension in albino CD-1 mice. Invest Ophthalmol Vis Sci 51:980–990

Fujikawa K, Iwata T, Inoue K, Akahori M, Kadotani H, Fukaya M, Watanabe M, Chang Q, Barnett EM, Swat W (2010) VAV2 and VAV3 as candidate disease genes for spontaneous glaucoma in mice and humans. PloS ONE 5:e9050

Gal A, Rau I, El Matri L, Kreienkamp HJ, Fehr S, Baklouti K, Chouchane I, Li Y, Rehbein M, Fuchs J, Fledelius HC, Vilhelmsen K, Schorderet DF, Munier FL, Ostergaard E, Thompson DA, Rosenberg T (2011) Autosomal-recessive posterior microphthalmos is caused by mutations in PRSS56, a gene encoding a trypsin-like serine protease. Am J Hum Genet 88:382–390

Gallenberger M, Kroeber M, Marz L, Koch M, Fuchshofer R, Braunger BM, Iwata T, Tamm ER (2014) Heterozygote Wdr36-deficient mice do not develop glaucoma. Exp Eye Res 128:83–91

Garcia-Valenzuela E, Shareef S, Walsh J, Sharma SC (1995) Programmed cell death of retinal ganglion cells during experimental glaucoma. Exp Eye Res 61:33–44

Gatzioufas Z, Hafezi F, Kopsidas K, Thanos S (2013) Dysfunctional uveoscleral pathway in a rat model of congenital glaucoma. J Physiol Pharmacol 64:393–397

Georgiou AL, Guo L, Francesca Cordeiro M, Salt TE (2014) Electroretinogram and visual-evoked potential assessment of retinal and central visual function in a rat ocular hypertension model of glaucoma. Cur Eye Res 39:472–486

Gharahkhani P, Burdon KP, Fogarty R, Sharma S, Hewitt AW, Martin S, Law MH, Cremin K, Bailey JN, Loomis SJ, Pasquale LR, Haines JL, Hauser MA, Viswanathan AC, McGuffin P, Topouzis F, Foster PJ, Graham SL, Casson RJ, Chehade M, White AJ, Zhou T, Souzeau E, Landers J, Fitzgerald JT, Klebe S, Ruddle JB, Goldberg I, Healey PR, Wellcome Trust Case Control C, Consortium N, Mills RA, Wang JJ, Montgomery GW, Martin NG, Radford-Smith G, Whiteman DC, Brown MA, Wiggs JL, Mackey DA, Mitchell P, MacGregor S, Craig JE (2014) Common variants near ABCA1, AFAP1 and GMDS confer risk of primary open-angle glaucoma. Nat Genet 46:1120–1125

Giovingo M, Nolan M, McCarty R, Pang IH, Clark AF, Beverley RM, Schwartz S, Stamer WD, Walker L, Grybauskas A, Skuran K, Kuprys PV, Yue BY, Knepper PA (2013) sCD44 overexpression increases intraocular pressure and aqueous outflow resistance. Mol Vis 19:2151–2164

Gobeil S, Rodrigue MA, Moisan S, Nguyen TD, Polansky JR, Morissette J, Raymond V (2004) Intracellular

sequestration of hetero-oligomers formed by wild-type and glaucoma-causing myocilin mutants. Invest Ophthalmol Vis Sci 45:3560–3567

Gould DB, Miceli-Libby L, Savinova OV, Torrado M, Tomarev SI, Smith RS, John SW (2004a) Genetically increasing Myoc expression supports a necessary pathologic role of abnormal proteins in glaucoma. Mol Cell Biol 24:9019–9025

Gould DB, Smith RS, John SW (2004b) Anterior segment development relevant to glaucoma. Int J Dev Biol 48:1015–1029

Gould DB, Reedy M, Wilson LA, Smith RS, Johnson RL, John SW (2006) Mutant myocilin nonsecretion in vivo is not sufficient to cause glaucoma. Mol Cell Biol 26:8427–8436

Grozdanic SD, Betts DM, Sakaguchi DS, Allbaugh RA, Kwon YH, Kardon RH (2003) Laser-induced mouse model of chronic ocular hypertension. Invest Ophthalmol Vis Sci 44:4337–4346

Guo Y, Cepurna WO, Dyck JA, Doser TA, Johnson EC, Morrison JC (2010) Retinal cell responses to elevated intraocular pressure: a gene array comparison between the whole retina and retinal ganglion cell layer. Invest Ophthalmol Vis Sci 51:3003–3018

Harada T, Harada C, Nakamura K, Quah HM, Okumura A, Namekata K, Saeki T, Aihara M, Yoshida H, Mitani A, Tanaka K (2007) The potential role of glutamate transporters in the pathogenesis of normal tension glaucoma. J Clin Invest 117:1763–1770

Harder JM, Fernandes KA, Libby RT (2012) The Bcl-2 family member BIM has multiple glaucoma-relevant functions in DBA/2J mice. Sci Rep 2:530

Harrison MM, Jenkins BV, O'Connor-Giles KM, Wildonger J (2014) A CRISPR view of development. Genes Dev 28:1859–1872

Hauser MA, Allingham RR, Linkroum K, Wang J, LaRocque-Abramson K, Figueiredo D, Santiago-Turla C, del Bono EA, Haines JL, Pericak-Vance MA, Wiggs JL (2006a) Distribution of WDR36 DNA sequence variants in patients with primary open-angle glaucoma. Invest Ophthalmol Vis Sci 47:2542–2546

Hauser MA, Sena DF, Flor J, Walter J, Auguste J, Larocque-Abramson K, Graham F, Delbono E, Haines JL, Pericak-Vance MA, Rand Allingham R, Wiggs JL (2006b) Distribution of optineurin sequence variations in an ethnically diverse population of low-tension glaucoma patients from the United States. J Glaucoma 15:358–363

He Y, Leung KW, Zhuo YH, Ge J (2009) Pro370Leu mutant myocilin impairs mitochondrial functions in human trabecular meshwork cells. Mol Vis 15:815–825

Howell GR, Libby RT, Jakobs TC, Smith RS, Phalan FC, Barter JW, Barbay JM, Marchant JK, Mahesh N, Porciatti V, Whitmore AV, Masland RH, John SW (2007) Axons of retinal ganglion cells are insulted in the optic nerve early in DBA/2J glaucoma. J Cell Biol 179:1523–1537

Howell GR, Libby RT, John SW (2008) Mouse genetic models: an ideal system for understanding glaucomatous neurodegeneration and neuroprotection. Prog Brain Res 173:303–321

Howell GR, Macalinao DG, Sousa GL, Walden M, Soto I, Kneeland SC, Barbay JM, King BL, Marchant JK, Hibbs M, Stevens B, Barres BA, Clark AF, Libby RT, John SW (2011) Molecular clustering identifies complement and endothelin induction as early events in a mouse model of glaucoma. J Clin Invest 121:1429–1444

Howell GR, Soto I, Zhu X, Ryan M, Macalinao DG, Sousa GL, Caddle LB, MacNicoll KH, Barbay JM, Porciatti V, Anderson MG, Smith RS, Clark AF, Libby RT, John SW (2012) Radiation treatment inhibits monocyte entry into the optic nerve head and prevents neuronal damage in a mouse model of glaucoma. J Clin Invest 122:1246–1261

Howell GR, Soto I, Ryan M, Graham LC, Smith RS, John SW (2013) Deficiency of complement component 5 ameliorates glaucoma in DBA/2J mice. J Neuroinflammation 10:76

Howell GR, MacNicoll KH, Braine CE, Soto I, Macalinao DG, Sousa GL, John SW (2014) Combinatorial targeting of early pathways profoundly inhibits neurodegeneration in a mouse model of glaucoma. Neurobiol Dis 71:44–52

Hysi PG, Cheng CY, Springelkamp H, Macgregor S, Bailey JN, Wojciechowski R, Vitart V, Nag A, Hewitt AW, Hohn R, Venturini C, Mirshahi A, Ramdas WD, Thorleifsson G, Vithana E, Khor CC, Stefansson AB, Liao J, Haines JL, Amin N, Wang YX, Wild PS, Ozel AB, Li JZ, Fleck BW, Zeller T, Staffieri SE, Teo YY, Cuellar-Partida G, Luo X, Allingham RR, Richards JE, Senft A, Karssen LC, Zheng Y, Bellenguez C, Xu L, Iglesias AI, Wilson JF, Kang JH, van Leeuwen EM, Jonsson V, Thorsteinsdottir U, Despriet DD, Ennis S, Moroi SE, Martin NG, Jansonius NM, Yazar S, Tai ES, Amouyel P, Kirwan J, van Koolwijk LM, Hauser MA, Jonasson F, Leo P, Loomis SJ, Fogarty R, Rivadeneira F, Kearns L, Lackner KJ, de Jong PT, Simpson CL, Pennell CE, Oostra BA, Uitterlinden AG, Saw SM, Lotery AJ, Bailey-Wilson JE, Hofman A, Vingerling JR, Maubaret C, Pfeiffer N, Wolfs RC, Lemij HG, Young TL, Pasquale LR, Delcourt C, Spector TD, Klaver CC, Small KS, Burdon KP, Stefansson K, Wong TY, Group BG, Consortium N, Wellcome Trust Case Control C, Viswanathan A, Mackey DA, Craig JE, Wiggs JL, van Duijn CM, Hammond CJ, Aung T (2014) Genome-wide analysis of multi-ancestry cohorts identifies new loci influencing intraocular pressure and susceptibility to glaucoma. Nat Genet 46:1126–1130

Inatani M, Tanihara H, Katsuta H, Honjo M, Kido N, Honda Y (2001) Transforming growth factor-beta 2 levels in aqueous humor of glaucomatous eyes. Graefe's Arch Clin Exp Ophthalmol 239:109–113

Iwase A, Suzuki Y, Araie M, Yamamoto T, Abe H, Shirato S, Kuwayama Y, Mishima HK, Shimizu H, Tomita G, Inoue Y, Kitazawa Y (2004) The prevalence of primary open-angle glaucoma in Japanese: the Tajimi Study. Ophthalmol 111:1641–1648

Jacobson N, Andrews M, Shepard AR, Nishimura D, Searby C, Fingert JH, Hageman G, Mullins R, Davidson BL, Kwon YH, Alward WL, Stone EM, Clark AF, Sheffield VC (2001) Non-secretion of mutant proteins of the glaucoma gene myocilin in cultured trabecular meshwork cells and in aqueous humor. Hum Mol Genet 10:117–125

Jakobs TC, Libby RT, Ben Y, John SW, Masland RH (2005) Retinal ganglion cell degeneration is topological but not cell type specific in DBA/2J mice. J Cell Biol 171:313–325

Joe MK, Tomarev SI (2010) Expression of myocilin mutants sensitizes cells to oxidative stress-induced apoptosis: implication for glaucoma pathogenesis. Am J Pathol 176:2880–2890

Joe MK, Sohn S, Hur W, Moon Y, Choi YR, Kee C (2003) Accumulation of mutant myocilins in ER leads to ER stress and potential cytotoxicity in human trabecular meshwork cells. Biochem Biophys Res Comm 312:592–600

Joe MK, Nakaya N, Abu-Asab M, Tomarev SI (2015) Mutated myocilin and heterozygous Sod2 deficiency act synergistically in a mouse model of open-angle glaucoma. Invest Ophthalmol Vis Sci 24:3322–3334

John SW, Harder JM, Fingert JH, Anderson MG (2014) Animal models of exfoliation syndrome, now and future. J Glaucoma 23:S68–72

Johnson TV, Tomarev SI (2010) Rodent models of glaucoma. Brain Res Bull 81:349–358

Johnson TV, Fan S, Toris CB (2008) Rebound tonometry in conscious, conditioned mice avoids the acute and profound effects of anesthesia on intraocular pressure. J Ocul Pharmacol Ther 24:175–185

Johnson EC, Cepurna WO, Choi D, Choe TE, Morrison JC (2014) Radiation pretreatment does not protect the rat optic nerve from elevated intraocular pressure-induced injury. Invest Ophthalmol Vis Sci 18:412–419

Junglas B, Kuespert S, Seleem AA, Struller T, Ullmann S, Bosl M, Bosserhoff A, Kostler J, Wagner R, Tamm ER, Fuchshofer R (2012) Connective tissue growth factor causes glaucoma by modifying the actin cytoskeleton of the trabecular meshwork. Am J Pathol 180:2386–2403

Kachaner D, Genin P, Laplantine E, Weil R (2012) Toward an integrative view of Optineurin functions. Cell Cycle 11:2808–2818

Kass MA, Heuer DK, Higginbotham EJ, Johnson CA, Keltner JL, Miller JP, Parrish RK, 2nd, Wilson MR, Gordon MO (2002) The ocular hypertension treatment study: a randomized trial determines that topical ocular hypotensive medication delays or prevents the onset of primary open-angle glaucoma. Arch Ophthalmol 120:701–713

Kim BS, Savinova OV, Reedy MV, Martin J, Lun Y, Gan L, Smith RS, Tomarev SI, John SW, Johnson RL (2001) Targeted disruption of the myocilin gene (myoc) suggests that human glaucoma-causing mutations are gain of function. Mol Cell Biol 21:7707–7713

Kim JH, Kang SY, Kim NR, Lee ES, Hong S, Seong GJ, Hong YJ, Kim CY (2011) Prevalence and characteristics of glaucoma among Korean adults. Korean J Ophthalmol 25:110–115

Kim TW, Kagemann L, Girard MJ, Strouthidis NG, Sung KR, Leung CK, Schuman JS, Wollstein G (2013) Imaging of the lamina cribrosa in glaucoma: perspectives of pathogenesis and clinical applications. Cur Eye Res 38:903–909

Kipfer-Kauer A, McKinnon SJ, Frueh BE, Goldblum D (2010) Distribution of amyloid precursor protein and amyloid-beta in ocular hypertensive C57BL/6 mouse eyes. Cur Eye Res 35:828–834

Kizhatil K, Ryan M, Marchant JK, Henrich S, John SW (2014) Schlemm's canal is a unique vessel with a combination of blood vascular and lymphatic phenotypes that forms by a novel developmental process. PLoS Biol 12:e1001912

Kotikoski H, Vapaatalo H, Oksala O (2003) Nitric oxide and cyclic GMP enhance aqueous humor outflow facility in rabbits. Cur Eye Res 26:119–123

Kumar S, Shah S, Deutsch ER, Tang HM, Danias J (2013) Triamcinolone acetonide decreases outflow facility in C57BL/6 mouse eyes. Invest Ophthalmol Vis Sci 54:1280–1287

Kwon HS, Nakaya N, Abu-Asab M, Kim HS, Tomarev SI (2014) Myocilin is involved in NgR1/Lingo-1-mediated oligodendrocyte differentiation and myelination of the optic nerve. J Neurosci 34:5539–5551

Levkovitch-Verbin H, Quigley HA, Martin KR, Valenta D, Baumrind LA, Pease ME (2002) Translimbal laser photocoagulation to the trabecular meshwork as a model of glaucoma in rats. Invest Ophthalmol Vis Sci 43:402–410

Li Y, Semaan SJ, Schlamp CL, Nickells RW (2007) Dominant inheritance of retinal ganglion cell resistance to optic nerve crush in mice. BMC Neurosci 8:19

Li X, McClellan ME, Tanito M, Garteiser P, Towner R, Bissig D, Berkowitz BA, Fliesler SJ, Woodruff ML, Fain GL, Birch DG, Khan MS, Ash JD, Elliott MH (2012) Loss of caveolin-1 impairs retinal function due to disturbance of subretinal microenvironment. J Biol Chem 287:16424–16434

Libby RT, Anderson MG, Pang IH, Robinson ZH, Savinova OV, Cosma IM, Snow A, Wilson LA, Smith RS, Clark AF, John SW (2005a) Inherited glaucoma in DBA/2J mice: pertinent disease features for studying the neurodegeneration. Vis Neurosci 22:637–648

Libby RT, Li Y, Savinova OV, Barter J, Smith RS, Nickells RW, John SW (2005b) Susceptibility to neurodegeneration in a glaucoma is modified by Bax gene dosage. PLoS Genet 1:17–26

Libby RT, Smith RS, Savinova OV, Zabaleta A, Martin JE, Gonzalez FJ, John SW (2003) Modification of ocular defects in mouse developmental glaucoma models by tyrosinase. Science 299:1578–1581

Liu Y, Vollrath D (2004) Reversal of mutant myocilin non-secretion and cell killing: implications for glaucoma. Hum Mol Genet 13:1193–1204

Liu Y, Allingham RR (2011) Molecular genetics in glaucoma. Exp Eye Res 93:331–339

Mabuchi F, Lindsey JD, Aihara M, Mackey MR, Weinreb RN (2004) Optic nerve damage in mice with a targeted type I collagen mutation. Invest Ophthalmol Vis Sci 45:1841–1845

McDowell CM, Luan T, Zhang Z, Putliwala T, Wordinger RJ, Millar JC, John SW, Pang IH, Clark AF (2012) Mutant human myocilin induces strain specific differences in ocular hypertension and optic nerve damage in mice. Exp Eye Res 100:65–72

McKinnon SJ, Schlamp CL, Nickells RW (2009) Mouse models of retinal ganglion cell death and glaucoma. Exp Eye Res 88:816–824

Mi XS, Zhang X, Feng Q, Lo AC, Chung SK, So KF (2012) Progressive retinal degeneration in transgenic mice with overexpression of endothelin-1 in vascular endothelial cells. Invest Ophthalmol Vis Sci 53:4842–4851

Minegishi Y, Iejima D, Kobayashi H, Chi ZL, Kawase K, Yamamoto T, Seki T, Yuasa S, Fukuda K, Iwata T (2013) Enhanced optineurin E50K-TBK1 interaction evokes protein insolubility and initiates familial primary open-angle glaucoma. Hum Mol Genet 22:3559–3567

Mitchell P, Smith W, Attebo K, Healey PR (1996) Prevalence of open-angle glaucoma in Australia. The Blue Mountains Eye Study. Ophthalmol 103:1661–1669

Mittag TW, Danias J, Pohorenec G, Yuan HM, Burakgazi E, Chalmers-Redman R, Podos SM, Tatton WG (2000) Retinal damage after 3 to 4 months of elevated intraocular pressure in a rat glaucoma model. Invest Ophthalmol Vis Sci 41:3451–3459

Monemi S, Spaeth G, DaSilva A, Popinchalk S, Ilitchev E, Liebmann J, Ritch R, Heon E, Crick RP, Child A, Sarfarazi M (2005) Identification of a novel adult-onset primary open-angle glaucoma (POAG) gene on 5q22.1. Hum Mol Genet 14:725–733

Morrison JC, Moore CG, Deppmeier LM, Gold BG, Meshul CK, Johnson EC (1997) A rat model of chronic pressure-induced optic nerve damage. Exp Eye Res 64:85–96

Morrison JC, Nylander KB, Lauer AK, Cepurna WO, Johnson E (1998) Glaucoma drops control intraocular pressure and protect optic nerves in a rat model of glaucoma. Invest Ophthalmol Vis Sci 39:526–531

Morrison JC, Jia L, Cepurna W, Guo Y, Johnson E (2009) Reliability and sensitivity of the TonoLab rebound tonometer in awake Brown Norway rats. Invest Ophthalmol Vis Sci 50:2802–2808

Nair KS, Hmani-Aifa M, Ali Z, Kearney AL, Ben Salem S, Macalinao DG, Cosma IM, Bouassida W, Hakim B, Benzina Z, Soto I, Soderkvist P, Howell GR, Smith RS, Ayadi H, John SW (2011) Alteration of the serine protease PRSS56 causes angle-closure glaucoma in mice and posterior microphthalmia in humans and mice. Nature Genet 43:579–584

Namekata K, Kimura A, Kawamura K, Guo X, Harada C, Tanaka K, Harada T (2013) Dock3 attenuates neural cell death due to NMDA neurotoxicity and oxidative stress in a mouse model of normal tension glaucoma. Cell Death Differ 20:1250–1256

Overby DR, Bertrand J, Tektas OY, Boussommier-Calleja A, Schicht M, Ethier CR, Woodward DF, Stamer WD, Lutjen-Drecoll E (2014) Ultrastructural changes associated with dexamethasone-induced ocular hypertension in mice. Invest Ophthalmol Vis Sci 55:4922–4933

Panagis L, Zhao X, Ge Y, Ren L, Mittag TW, Danias J (2010) Gene expression changes in areas of focal loss of retinal ganglion cells in the retina of DBA/2J mice. Invest Ophthalmol Vis Sci 51:2024–2034

Panagis L, Zhao X, Ge Y, Ren L, Mittag TW, Danias J (2011) Retinal gene expression changes related to IOP exposure and axonal loss in DBA/2J mice. Invest Ophthalmol Vis Sci 52:7807–7816

Pang IH, Clark AF (2007) Rodent models for glaucoma retinopathy and optic neuropathy. J Glaucoma 16:483–505

Park DY, Lee J, Park I, Choi D, Lee S, Song S, Hwang Y, Hong KY, Nakaoka Y, Makinen T, Kim P, Alitalo K, Hong YK, Koh GY (2014) Lymphatic regulator PROX1 determines Schlemm's canal integrity and identity. J Clin Invest 124:3960–3974

Pease ME, Cone FE, Gelman S, Son JL, Quigley HA (2011) Calibration of the TonoLab tonometer in mice with spontaneous or experimental glaucoma. Invest Ophthalmol Vis Sci 52:858–864

Prasanna G, Krishnamoorthy R, Yorio T (2011) Endothelin, astrocytes and glaucoma. Exp Eye Res 93:170–177

Quigley HA, Broman AT (2006) The number of people with glaucoma worldwide in 2010 and 2020. British J Ophthalmol 90:262–267

Rao KN, Kaur I, Parikh RS, Mandal AK, Chandrasekhar G, Thomas R, Chakrabarti S (2010) Variations in NTF4, VAV2, and VAV3 genes are not involved with primary open-angle and primary angle-closure glaucomas in an indian population. Invest Ophthalmol Vis Sci 51:4937–4941

Raymond ID, Pool AL, Vila A, Brecha NC (2009) A Thy1-CFP DBA/2J mouse line with cyan fluorescent protein expression in retinal ganglion cells. Vis Neurosci 26:453–465

Rezaie T, Child A, Hitchings R, Brice G, Miller L, Coca-Prados M, Heon E, Krupin T, Ritch R, Kreutzer D, Crick RP, Sarfarazi M (2002) Adult-onset primary open-angle glaucoma caused by mutations in optineurin. Science 295:1077–1079

Rosen AM, Stevens B (2010) The role of the classical complement cascade in synapse loss during development and glaucoma. Adv Exp Med Biol 703:75–93

Ruiz-Ederra J, Verkman AS (2006) Mouse model of sustained elevation in intraocular pressure produced by episcleral vein occlusion. Exp Eye Res 82:879–884

Sappington RM, Carlson BJ, Crish SD, Calkins DJ (2010) The microbead occlusion model: a paradigm for induced ocular hypertension in rats and mice. Invest Ophthalmol Vis Sci 51:207–216

Senatorov V, Malyukova I, Fariss R, Wawrousek EF, Swaminathan S, Sharan SK, Tomarev S (2006) Expression of mutated mouse myocilin induces open-angle glaucoma in transgenic mice. J Neuroscie 26:11903–11914

Seo S, Solivan-Timpe F, Roos BR, Robin AL, Stone EM, Kwon YH, Alward WL, Fingert JH (2013) Identification of proteins that interact with TANK binding kinase 1 and testing for mutations associated with glaucoma. Cur Eye Res 38:310–315

Shareef SR, Garcia-Valenzuela E, Salierno A, Walsh J, Sharma SC (1995) Chronic ocular hypertension following episcleral venous occlusion in rats. Exp Eye Res 61:379–382

Shepard AR, Millar JC, Pang IH, Jacobson N, Wang WH, Clark AF (2010) Adenoviral gene transfer of active human transforming growth factor-{beta}2 elevates intraocular pressure and reduces outflow facility in rodent eyes. Invest Ophthalmol Vis Sci 51:2067–2076

Shi D, Takano Y, Nakazawa T, Mengkegale M, Yokokura S, Nishida K, Fuse N (2013) Molecular genetic analysis of primary open-angle glaucoma, normal tension glaucoma, and developmental glaucoma for the VAV2 and VAV3 gene variants in Japanese subjects. Biochem Biophys Res Com 432:509–512

Smedowski A, Pietrucha-Dutczak M, Kaarniranta K, Lewin-Kowalik J (2014) A rat experimental model of glaucoma incorporating rapid-onset elevation of intraocular pressure. Sci Rep 4:5910

Sommer A, Tielsch JM, Katz J, Quigley HA, Gottsch JD, Javitt J, Singh K (1991) Relationship between intraocular pressure and primary open angle glaucoma among white and black Americans. The Baltimore Eye Survey. Arch Ophthalmol 109:1090–1095

Soto I, Pease ME, Son JL, Shi X, Quigley HA, Marsh-Armstrong N (2011) Retinal ganglion cell loss in a rat ocular hypertension model is sectorial and involves early optic nerve axon loss. Invest Ophthalmol Vis Sci 52:434–441

Sowden JC (2007) Molecular and developmental mechanisms of anterior segment dysgenesis. Eye 21:1310–1318

Stasi K, Nagel D, Yang X, Wang RF, Ren L, Podos SM, Mittag T, Danias J (2006) Complement component 1Q (C1Q) upregulation in retina of murine, primate, and human glaucomatous eyes. Invest Ophthalmol Vis Sci 47:1024–1029

Steele MR, Inman DM, Calkins DJ, Horner PJ, Vetter ML (2006) Microarray analysis of retinal gene expression in the DBA/2J model of glaucoma. Invest Ophthalmol Vis Sci 47:977–985

Steinhart MR, Cone FE, Nguyen C, Nguyen TD, Pease ME, Puk O, Graw J, Oglesby EN, Quigley HA (2012) Mice with an induced mutation in collagen 8A2 develop larger eyes and are resistant to retinal ganglion cell damage in an experimental glaucoma model. Mol Vis 18:1093–1106

Steinhart MR, Cone-Kimball E, Nguyen C, Nguyen TD, Pease ME, Chakravarti S, Oglesby EN, Quigley HA (2014) Susceptibility to glaucoma damage related to age and connective tissue mutations in mice. Exp Eye Res 119:54–60

Stevens B, Allen NJ, Vazquez LE, Howell GR, Christopherson KS, Nouri N, Micheva KD, Mehalow AK, Huberman AD, Stafford B, Sher A, Litke AM, Lambris JD, Smith SJ, John SW, Barres BA (2007) The classical complement cascade mediates CNS synapse elimination. Cell 131:1164–1178

Stewart WC, Magrath GN, Demos CM, Nelson LA, Stewart JA (2011) Predictive value of the efficacy of glaucoma medications in animal models: preclinical to regulatory studies. British J Ophthalmol 95:1355–1360

Stoilov I, Akarsu AN, Sarfarazi M (1997) Identification of three different truncating mutations in cytochrome P4501B1 (CYP1B1) as the principal cause of primary congenital glaucoma (Buphthalmos) in families linked to the GLC3A locus on chromosome 2p21. Hum Mol Genet 6:641–647

Stone EM, Fingert JH, Alward WL, Nguyen TD, Polansky JR, Sunden SL, Nishimura D, Clark AF, Nystuen A, Nichols BE, Mackey DA, Ritch R, Kalenak JW, Craven ER, Sheffield VC (1997) Identification of a gene that causes primary open angle glaucoma. Science 275:668–670

Teixeira LB, Zhao Y, Dubielzig RR, Sorenson CM, Sheibani N (2014) Ultrastructural abnormalities of the trabecular meshwork extracellular matrix in Cyp1b1-deficient mice. Vet Pathol 52:397–403

Tham YC, Li X, Wong TY, Quigley HA, Aung T, Cheng CY (2014) Global prevalence of glaucoma and projections of glaucoma burden through 2040: asystematic review and meta-analysis. Ophthalmology 121:2081–2090

Thanos S, Naskar R (2004) Correlation between retinal ganglion cell death and chronically developing inherited glaucoma in a new rat mutant. Exp Eye Res 79:119–129

Thorleifsson G, Magnusson KP, Sulem P, Walters GB, Gudbjartsson DF, Stefansson H, Jonsson T, Jonasdottir A, Jonasdottir A, Stefansdottir G, Masson G, Hardarson GA, Petursson H, Arnarsson A, Motallebipour M, Wallerman O, Wadelius C, Gulcher JR, Thorsteinsdottir U, Kong A, Jonasson F, Stefansson K (2007) Common sequence variants in the LOXL1 gene confer susceptibility to exfoliation glaucoma. Science 317:1397–1400

Tomarev SI, Wistow G, Raymond V, Dubois S, Malyukova I (2003) Gene expression profile of the human trabecular meshwork: NEIBank sequence tag analysis. Invest Ophthalmol Vis Sci 44:2588–2596

Trantow CM, Mao M, Petersen GE, Alward EM, Alward WL, Fingert JH, Anderson MG (2009) Lyst mutation in mice recapitulates iris defects of human exfoliation syndrome. Invest Ophthalmol Vis Sci 50:1205–1214

Tripathi RC, Li J, Chan WF, Tripathi BJ (1994) Aqueous humor in glaucomatous eyes contains an increased level of TGF-beta 2. Exp Eye Res 59:723–727

Tsuruga H, Murata H, Araie M, Aihara M (2012) A model for the easy assessment of pressure-dependent damage to retinal ganglion cells using cyan fluorescent protein-expressing transgenic mice. Mol Vis 18:2468–2478

Ueda J, Sawaguchi S, Hanyu T, Yaoeda K, Fukuchi T, Abe H, Ozawa H (1998) Experimental glaucoma model in the rat induced by laser trabecular photocoagulation after an intracameral injection of India ink. Jpn J Ophthalmol 42:337–344

Urcola JH, Hernandez M, Vecino E (2006) Three experimental glaucoma models in rats: comparison of the effects of intraocular pressure elevation on retinal ganglion cell size and death. Exp Eye Res 83:429–437

Vasiliou V, Gonzalez FJ (2008) Role of CYP1B1 in glaucoma. Ann Rev Pharmacol Toxicol 48:333–358

Wiggs JL, Pawlyk B, Connolly E, Adamian M, Miller JW, Pasquale LR, Haddadin RI, Grosskreutz CL, Rhee DJ, Li T (2014) Disruption of the blood-aqueous barrier and lens abnormalities in mice lacking lysyl oxidase-like 1 (LOXL1). Invest Ophthalmol Vis Sci 55:856–864

Williams PA, Howell GR, Barbay JM, Braine CE, Sousa GL, John SW, Morgan JE (2013) Retinal ganglion cell dendritic atrophy in DBA/2J glaucoma. PloS ONE 8:e72282

Yang Q, Cho KS, Chen H, Yu D, Wang WH, Luo G, Pang IH, Guo W, Chen DF (2012) Microbead-induced ocular hypertensive mouse model for screening and testing of aqueous production suppressants for glaucoma. Invest Ophthalmol Vis Sci 53:3733–3741

Yun H, Lathrop KL, Yang E, Sun M, Kagemann L, Fu V, Stolz DB, Schuman JS, Du Y (2014) A laser-induced mouse model with long-term intraocular pressure elevation. PloS ONE 9:e107446

Zhang Z, Dhaliwal AS, Tseng H, Kim JD, Schuman JS, Weinreb RN, Loewen NA (2014) Outflow tract ablation using a conditionally cytotoxic feline immunodeficiency viral vector. Invest Ophthalmol Vis Sci 55:935–940

Zhao Y, Wang S, Sorenson CM, Teixeira L, Dubielzig RR, Peters DM, Conway SJ, Jefcoate CR, Sheibani N (2013) Cyp1b1 mediates periostin regulation of trabecular meshwork development by suppression of oxidative stress. Mol Cell Biol 33:4225–4240

Zhou Y, Grinchuk O, Tomarev SI (2008) Transgenic mice expressing the Tyr437His mutant of human myocilin protein develop glaucoma. Invest Ophthalmol Vis Sci 49:1932–1939

Zode GS, Kuehn MH, Nishimura DY, Searby CC, Mohan K, Grozdanic SD, Bugge K, Anderson MG, Clark AF, Stone EM, Sheffield VC (2011) Reduction of ER stress via a chemical chaperone prevents disease phenotypes in a mouse model of primary open angle glaucoma. J Clin Invest 121:3542–3553

Zode GS, Sharma AB, Lin X, Searby CC, Bugge K, Kim GH, Clark AF, Sheffield VC (2014) Ocular-specific ER stress reduction rescues glaucoma in murine glucocorticoid-induced glaucoma. J Clin Invest 124:1956–1965

Commentary

John C. Morrison
e-mail: morrisoj@ohsu.edu
Casey Eye Institute,
Oregon Health and Science University,
Portland, OR, USA

Lowering intraocular pressure (IOP) remains our best treatment for preserving vision in glaucoma patients, and all therapies—medical, laser, and surgical—are directed at achieving this goal. For practicing clinicians, one of the most difficult challenges is posed by the glaucoma patient who demonstrates progressive visual field loss despite these treatments. This may result from poor response or tolerance to medical treatment, failure of surgery to adequately lower IOP, or situations where the surgery cannot be performed due to an unacceptable risk of surgical complications. Still in others, visual field loss progresses even though the pressure has been brought well into the single digits. In all of these cases, IOP simply remains higher than the threshold of injury. This means that our best hope in these difficult situations is the potential of someday developing neuroprotective therapies that will help the glaucomatous optic nerve, and retina, better withstand the effects of IOP. This depends on understanding cellular mechanisms involved in pressure-induced axonal injury.

While primate models, once considered essential for glaucoma research, possess optic nerve head (ONH) anatomy identical to the human, less expensive and more available rodent models offer excellent alternatives for understanding how elevation and fluctuation of IOP lead to axonal injury. This is because both the rat and mouse possess the essential cellular relationships that make the ONH, the primary site of injury, unique.

Accompanied by a delicate capillary bed derived from retinal and posterior ciliary sources, unmyelinated axon bundles from the nerve fiber layer turn posteriorly to form the optic nerve.

Here, they are separated and supported by ONH astrocytes oriented across the scleral canal, lying perpendicular to the axon bundles. These astrocytes form glial columns, analogous to collagenous laminar beams of the primate ONH, and send numerous processes into the nerve bundles to provide intimate axonal support. Thus, the astrocytes of the rodent ONH are perfectly oriented to detect and respond to stresses and strains induced by changing IOP, and in turn affect axon function and viability.

This excellent review by Johnson and Tomarev thoroughly summarizes the literal explosion in rodent glaucoma models that has occurred over the past 20 years. These range from induced methods that rely on deliberate obstruction of aqueous humor outflow, which have the advantages of being unilateral with a tightly controlled onset, to genetic models, in which IOP elevation tends to be more modest and potentially more like that in human glaucoma. Many of the latter are based on genes associated with human glaucoma, offering opportunities to study mechanisms that lead to elevated IOP, in addition to understanding the consequences to the optic nerve. Coincident with the development of these models, modern cell biology methods are now available which allow us to study IOP-induced gene expression and protein changes in these small nerve heads. There is every reason to believe that cellular events uncovered by studying these models are also active in human glaucoma and that, with all of these tools, they will ultimately lead to effective neuroprotective strategies for our most vulnerable patients.

Animal Models of Age-Related Macular Degeneration: Subretinal Inflammation

4

Florian Sennlaub

4.1 Introduction

Ideally, an animal model of a given human disease should be triggered by the same risk or causative factors observed in humans and lead to a similar pathology. This is, to a large degree, the case in monogenic disease, when transgenic animals bearing the human genetic mutation (and not the knockout of the whole corresponding gene) develop a similar pathology observed in humans. We could call these models "first-degree" models as they rather precisely reproduce the origin and ensuing pathomechanism of a given disease. These are ideal models to decipher how the causative mutation alters molecular and cellular homeostasis and leads to tissue alterations and disease. Most diseases, however, are complex and caused due to several factors that are often not all known. In these cases the chronic interplay of several risk/causative factors leads to secondary and tertiary events such as the death of certain cell population, accumulation of waste products and debris, inflammation, and/or the

uncontrolled proliferation of certain cell types (endothelial cells, fibrous tissue). Animal models can be designed to mimic certain downstream stages of the disease and decipher subsequent mechanisms. "Downstream" models usually reproduce only one or a limited number of phenotypes/symptoms of the complex disease which are often not specific to the complex disease. They are by definition more problematic, as they induce a phenotypically similar phenotype in animals by other means than the risk/causative factors that are responsible for the human disease. They can be useful to understand downstream mechanisms that induce the similar phenotype but have to be interpreted with caution as distinct mechanisms might lead to the same phenotype.

Age-related macular degeneration (AMD) is a common, complex disease that results from interplay of genetic and environmental risk factors. Polymorphisms in the complement factor H (CFH, an inhibitor of the alternative complement cascade), the promotor of the high-temperature requirement A serine peptidase 1 (HTRA1, a serine protease involved in inflammation), and the apolipoprotein E isoforms (ApoE, a lipoprotein involved in reverse cholesterol transport and immune regulation) likely account for a large part of the genetic risk to AMD (Swaroop et al. 2007). Age and smoking are the main environmental risk factors for AMD, but other factors such as poor diet and excessive light exposure likely play a role (Chakravarthy et al. 2010). Given the multifactorial nature of the disease and that none of the risk factors on its own is sufficient to trigger

F. Sennlaub (✉)
Inserm, U 968, UPMC University Paris 06,
75012 Paris, France
e-mail: florian.sennlaub@inserm.fr

UMR_S 968, Institut de la Vision, UPMC University
Paris 06, 75012 Paris, France

INSERM-DHOS CIC 503, Centre Hospitalier National
d'Ophtalmologie des Quinze-Vingts,
75012 Paris, France

the disease in humans, it is to be expected that animal models mimicking a single factor, or a combination of only two risk factors, is not sufficient to produce all aspects of AMD. However, these models might be able to reproduce different features of AMD and provide an entry point into the understanding of the pathomechanism and the identification of potential drug targets.

Subretinal inflammation is observed in early/ intermidate, and late AMD (see below). It is clearly not specific to AMD, but this is also true for drusen, photoreceptor/retinal pigment epithelium (RPE) cell death, and choroidal neovascularization (CNV) as they are all individually observed in other diseases and are the manifestations of ill-defined dysfunctions. We will also see in this chapter that some "first-degree" animal models of AMD-risk factors suggest that impaired subretinal immune suppression and inflammation is one of the initiating events in AMD. Primary or secondary, non-resolving, chronic inflammation often becomes pathogenic and contributes to disease progression, as seen in many chronic inflammatory diseases (Nathan and Ding 2010). To understand the initiation of chronic subretinal inflammation and its possible pathogenic role in AMD, we are greatly dependent on animal models. In this chapter we first summarize the evidence and nature of subretinal inflammation in AMD. We next review whether and how known AMD risk factors influence eventual subretinal inflammation in "first-degree" AMD animal models. We next outline "primary" inflammation models, where the subretinal inflammation is due to a defect in tonic anti-inflammatory signals, deficient leukocyte elimination, or an autoimmune reaction. Last but not least, we give an overview on "secondary" inflammation models, where the subretinal inflammation occurs after a primary injury or degeneration. We concentrate on mouse models, as the vast majority of research in the field is currently carried out on rodent models.

4.2   Inflammation and Age-Related Macular Degeneration

Inflammation is a frequent occurrence in the context of tissue injury (sterile inflammation) and microbial infection. Mononuclear phagocytes (MPs) play a central role in the inflammatory reaction. MPs comprise a family of cells that include monocytes (Mos), inflammatory and resident macrophages (iMφ, rMφ), microglial cells (MC, the rMφ of the central nervous system) and dendritic cells (DC; Gautier et al. 2012). MPs survey the local tissue environment, and are involved in the innate and adaptive immune responses that occur during inflammation. Inflammation is typically characterized by rMφ/MC activation, early neutrophil recruitment followed by an influx of Mos that differentiate into iMφ (Wynn et al. 2013; Nathan and Ding 2010). They secrete a variety of inflammatory mediators which activate antimicrobial defense mechanisms, including oxidative processes and complement activation that contribute to the killing of invading organisms (Wynn et al. 2013). Some iMφs and inflammatory DCs can migrate from the injury site to the draining lymph nodes (LNs), where they initiate an adaptive immune response. At the injury site, neutrophils undergo death within hours and are cleared together with pathogens and tissue debris by macrophages. Phagocytosis triggers the production of mediators which generate macrophages that facilitate tissue repair and/or scar formation and resolution (Fadok et al. 2001; Huynh et al. 2002). Finally, iMφs disappear from the site and the tissue is left with the tissue-specific rMφ (Gautier et al. 2013). Ideally, inflammation quickly and efficiently eliminates pathogens and repairs the tissue injury either by regeneration or by scarring. If the inflammatory response is not quickly controlled, it can become pathogenic and contribute to disease progression, as seen in many chronic inflammatory diseases (Nathan and Ding 2010).

The inflammatory response consists of a compromise of fast effective elimination and neutralization of pathogens, and avoidance of collateral tissue damage that interferes with vital functions. The retina is especially vulnerable to immunopathic damage as it has very limited regenerative capacities. A blood–tissue barrier and a physiological immunosuppressive environment particularly protect it from excessive leukocyte infiltration. Thereby, innate and adaptive immune responses are dampened or suppressed, possibly to minimize collateral damage of the inflammatory response (Streilein 2003).

Physiologically, the posterior segment of the eye contains numerous rMφs in the choroid and ciliary body and some perivascular rMφs in the retina. A network of MCs is located in the inner retinal layers, but a number of neuron- and glial-derived factors continually represses their activation (e.g., CX3CL1, CD200L, SIRP 1α, CD22; Galea et al. 2007). The central subretinal space, located between the RPE and the photoreceptor outer segments (POS), is physiologically devoid of MPs in healthy subjects (Gupta et al. 2003; Combadiere et al. 2007; Sennlaub et al. 2013), and they are only present in the extreme periphery (McMenamin and Loeffler 1990). The immunosuppressive capacities of the healthy RPE likely contribute to this peculiarity: the RPE induces apoptosis of lymphocytes and iMφs efficiently. The immunosuppressivity of the RPE is well illustrated by the fact that RPE allografts to non-immune privileged sites of non-immuno-compatible hosts can survive prolonged periods of time (Wenkel and Streilein 2000). Furthermore, the retina has no lymphatic drainage system and only a few DCs (Streilein 2003; Forrester et al. 2010; Kumar et al. 2014), and subretinal space lacks blood vessels through which effector cells could infiltrate the tissue (Streilein et al. 2002).

Early and intermediate AMD is characterized by sizeable deposits of lipoproteinaceous debris, called large drusen, located in Bruch's membrane (BM) and partially covered by the RPE (Sarks 1976). The presence of large drusen is an important risk factor for late AMD (Klein et al. 2004). There are two clinical forms of late AMD: wet AMD, which is defined by CNV, and geographic atrophy (GA), which is characterized by an extending lesion of both the RPE and photoreceptors (Sarks 1976).

The first histological evidence of subretinal leucocytes in exudative AMD dates back to 1916 (Hegner 1916) and has been reproduced in numerous publications since (Penfold et al. 2001). In particular MPs and lymphocytes are invariably observed in excised neovascular membranes from patients with AMD (for review see (Penfold et al. 2001)). In GA, MPs have been observed within the atrophic area (Gupta et al. 2003; Combadiere et al. 2007; Penfold et al. 2001; Levy et al. 2015), on RPE cells adjacent to the lesions (Gupta et al. 2003; Sennlaub et al. 2013; Levy et al. 2015), and within and around drusen (Sennlaub et al. 2013; Levy et al. 2015; Killingsworth et al. 1990; Hageman et al. 2001).

Most of the published studies concerning MP accumulation in AMD do not permit the differentiation of distinct types of MPs, such as Mo-derived iMφs and MCs, as they are morphologically indistinguishable and express or induce similar markers (Gautier et al. 2012; Ransohoff and Cardona 2010). Using CCR2 as a specific marker, of iMos/iMφs (Geissmann et al. 2010), as it not expressed and cannot be induced in MCs (Saederup et al. 2010), we have recently shown that inflammatory CCR2+ Mos/iMφs invariably take part in the subretinal inflammation in GA lesions and large drusen (Sennlaub et al. 2013).

Lymphocytes are rarely observed in the retina of AMD patients, but infiltrate the choroid and neovascular membranes (Penfold et al. 2001). Furthermore, interleukin (IL) 17A, a pro-inflammatory cytokine and major product of CD4+ T helper 17 lymphocytes is increased in eyes with GA (Wei et al. 2012). A recent report also shows an increased number of degranulated mast cells in the choroid of GA patients (Lutty et al. 2013).

The inflammation in AMD is relatively mild and a neutrophil infiltration has not been described, compared to fast evolving autoimmune retinitis lesions, in which an invasion of cytotoxic T lymphocytes, neutrophils, and macrophages are observed (Kerr et al. 2008). This difference might reflect the slow progression of the degeneration bserved in GA compared to fast tissue destruction that characterizes chorioretinitis.

To date, it is not clear if the observed inflammatory changes are purely a result of degenerative and neovascular changes and do not influence yje disease, if they initiate AMD or participate in AMD progression, or if they are protective for the tissue. The observation that chorioretinal inflammation can be sufficient to cause CNV in young patients with chorioretinits in the absence of prior degenerative changes, and that immunosuppressive agents in wet AMD patients lower the need for anti-vascular endothelial growth factor (VEGF) therapy (Nussenblatt et al. 2010) suggests that the inflammation can be pathogenic at least in exudative AMD (for a review of anti-inflammatory agents in AMD see Wang et al. (Wang et al. 2011)).

Animal models of subretinal inflammation can help to further elucidate whether and how AMD risk factors influence subretinal inflammation, and whether subretinal inflammation participates in CNV and degeneration.

4.3 Subretinal Inflammation in AMD Animal Models

4.3.1 AMD-Associated Risk Factors and Inflammation

4.3.1.1 Genetic Risk Factors

The common polymorphisms in CFH, HTRA1, and the ApoE isoforms account for a large part of the genetic risk to late AMD (Swaroop et al. 2007). They are also associated with early/intermediate AMD (Yu et al. 2012), which might signify that they are not specific to RPE/photoreceptor degeneration or CNV, but implicated in pathomechanisms that are common to early/intermediate and late AMD. Such a common mechanism could be subretinal inflammation but it is not yet clear whether or how the disease-associated alleles influence subretinal inflammation. Interestingly, all three proteins are expressed at high levels in activated MPs (HTRA1 (Hou et al. 2013), ApoE (Levy et al. 2015; Peri and Nusslein-Volhard 2008; Basu et al. 1982; Rosenfeld et al. 1993), and CFH (Griffiths et al. 2009; Calippe et al. 2014)) and might be directly involved in MP function.

A common polymorphism or isoform appears in a population, when a germ line mutation in the ancestral version of an allele generates an advantage in the "fitness" (survival, reproduction, etc.) of its carrier. The frequency of the allele thereby increases over the generations in the population under evolutionary pressure. Infectious diseases exert a constant, strong evolutionary pressure on the genetic makeup of a population. The AMD-associated polymorphisms might have been selected because they lead to a stronger inflammatory response and better defense against infectious disease.

ApoE, for example, plays a role in susceptibility to infection (Roselaar and Daugherty 1998; de Bont et al. 1999; Sinnis et al. 1996) and *APOE3* allele carriers are somewhat protected against progression and death from HIV infection (Burt et al. 2008) compared with the *APOE4* genotype, believed to be the ancestral isoform (Raichlen and Alexander 2014). On the other hand, *APOE3*-allele carriers carry an increased risk for AMD compared with *APOE4*-allele carriers, which further increases in *APOE2*-allele carriers (McKay et al. 2011). The *APOE2* isoform is associated with increased ApoE transcription and protein levels in humans and in humanized transgenic mice that express the *APOE2* allele (Yu et al. 2013; Mahley and Rall 2000). In mice, we have recently shown that ApoE overexpression in MPs, observed in Cx3cr1-deficient mice, downregulates RPE FasL expression via IL-6 and leads to prolonged survival and accumulation of subretinal MPs (Levy et al. 2015). Overexpression of ApoE in *APOE2* mice might lead to a similar effect and promote subretinal MP accumulation. *APOE2* mice might present a "first-degree" AMD model for subretinal inflammation in the future.

Similarly, the AMD-associated 402H CFH, which alters CFH surface binding depending on the nature of the surface (Clark et al. 2006), binds significantly less to certain pathogenic bacteria that have evolved the capacity to bind CFH and escape complement-mediated elimination (Haapasalo et al. 2008). The 402H allele was proposed to have emerged in the European population because it was a survival factor during bacterial pandemics (Clark et al. 2010). To our knowledge, most studies that analyzed CFH location in AMD have shown that CFH is increased in the choroid and in drusen of AMD patients (for a review see Calippe et al. (Calippe et al. 2014)). Interestingly, the transgenic expression of CFH under the APOE promoter, which likely leads to increased CFH expression in the eye where ApoE is strongly transcribed, develops significant age-dependent subretinal inflammation compared with control mice, suggesting that CFH is involved in subretinal MP accumulation (Ufret-Vincenty et al. 2012). However the increased accumulation was independent of the polymorphism and better models, such as CFH H knockin mice, are needed that precisely replicate the human risk allele to determine whether and how the CFH polymorphism directly affects subretinal inflammation.

Last but not least, HTRA1, which promotes inflammation (Grau et al. 2006), is increased in carriers of the AMD-associated single nucleotide polymorphism (SNP; Yang et al. 2006; Chan et al. 2007), and transgenic mice that overexpress HTRA1 have been reported to present a subretinal MP accumulation (Liao et al. 2013).

Taken together, the three main AMD-associated alleles affect proteins involved in inflammation and might have been selected in certain populations due to the evolutionary pressure of infectious diseases. Animal models that more or less accurately mimic the AMD-associated allele or allele effect appear to promote subretinal inflammation. It will be interesting to see whether humanized knockin mice that express the AMD risk alleles develop a similar phenotype.

4.3.1.2   AMD-Associated Environmental Risk Factors and Inflammation

Age and smoking are the main environmental risk factors for AMD (Chakravarthy et al. 2010), but other factors such as the patient's pigmentation and excessive light exposure might also play a role (Klein et al. 2014). Age has been shown to be sufficient to induce subretinal accumulation of MPs in wild-type (WT) mice (Xu et al. 2008). Similarly, conventional, nontoxic light (250 lux) is enough to induce subretinal MP accumulation in albino mouse strains, but not in pigmented mice (Ng and Streilein 2001). The fact that subretinal MPs do not accumulate when the albino mice are raised in darkness and are eliminated when placed into darkness clearly shows the light dependence of the accumulation in these albino mouse strains. Age and/or light exposure are also essential in the accumulation of subretinal MPs in "primary inflammation" models (see below). To our knowledge there are no studies investigating a subretinal inflammation in experimental chronic cigarette smoke exposure to date, but smoking has been shown to exacerbate a variety

of auto-inflammatory and autoimmune diseases (Arnson et al. 2010).

Taken together, several animal models that mimic individual AMD risk factors are sufficient to induce subretinal MP accumulation but not degeneration or CNV, which might suggest that subretinal inflammation is an early feature in the pathogenesis. It remains to be seen whether one model that unites all known genetic and environmental risk factors leads to more severe subretinal inflammation that could trigger degeneration and CNV and constitute an ideal model of AMD.

4.3.2   "Primary inflammation" AMD Models

Several animal models of subretinal inflammation exist, in which the inflammation is not a secondary event to photoreceptor or RPE dysfunction but due to (i) deficient physiological tonic anti-inflammatory signals, (ii) deficient immunosuppressive environment, or (iii) the induction of an autoimmune response to subretinal epitopes.

4.3.2.1   Suppression of Tonic Anti-Inflammatory Signals

Physiologically, neurons express a number of factors that continually repress MC activation, such as CX3CL1, CD200L, SIRP 1α, and CD22 (Galea et al. 2007). In the retina, CX3CL1 is constitutively expressed as a transmembrane protein in inner retinal neurons (Zieger et al. 2014) and provides a tonic inhibitory signal to CX3CR1-expressing retinal MCs that keeps these cells in a quiescent surveillance mode under physiological conditions (Ransohoff 2009). Deletion of Cx3cr1 leads to an accelerated age-dependent increase of subretinal MPs in pigmented animals in both Cx3cr1$^{-/-}$ knockout (Combadiere et al. 2007) and Cx3cr1$^{GFP/GFP}$ knockin mice (Combadiere et al. 2007; Sennlaub et al. 2013; Levy et al. 2015; Chinnery et al. 2011) compared with WT animals kept under the same light conditions (~250 lux). The phenotype is not due to the Crb1^{rd8} contamination (see below) as the mice were not contaminated in these experiments. Cx3cr1 deletion also leads to a significant accumulation in young albino mice (Combadiere et al. 2007; Chinnery et al. 2011) and a light challenge (that is not itself toxic, as it does not induce subretinal inflammation or degeneration in control mice) induces MP accumulation in pigmented Cx3cr1$^{GFP/GFP}$ mice (Sennlaub et al. 2013; Levy et al. 2015). Raising albino Cx3cr1$^{-/-}$ mice in darkness (Combadiere et al. 2007) or letting pigmented C57BL/6 Cx3cr1$^{GFP/GFP}$ mice age in very dim light conditions prevents the accumulation (Luhmann et al. 2013; Combadiere et al. 2013), which confirms the importance of light in subretinal MP accumulation (Ng and Streilein 2001). Strikingly similar to CX3CL1, CD200 is expressed by neurons of the inner retina and provides a tonic inhibitory signal via the CD200R receptor expressed by MPs (Broderick et al. 2002). Cx3cr1$^{-/-}$ and CD200R$^{-/-}$ mice develop exaggerated subretinal MP accumulation and CNV (Combadiere et al. 2007; Horie et al. 2013). Furthermore, the age-, light-, and laser-dependent accumulation of subretinal Cx3cr1-deficient MPs is associated with a significant loss in photoreceptors (Combadiere et al. 2007; Chinnery et al. 2011). Mechanistically, we showed that the lack of the tonic inhibitory signal of CX3CR1 leads to increased expression of CCL2 and recruitment of neurotoxic blood-derived iMφs under the retina, also observed in GA patients (Sennlaub et al. 2013). Furthermore, Cx3cr1-deficient MPs of all origins studied (peritoneal Mφs, bone marrow derived Mos, and brain MCs) are more resistant to clearance after subretinal adoptive transfer to WT recipients (Levy et al. 2015). We showed that APOE-induced IL-6 release from the MPs represses RPE FasL-mediated immune-suppression, and prolongs subretinal MP survival (Levy et al. 2015). Taken together the increased survival (Levy et al. 2015) and recruitment (Sennlaub et al. 2013) likely explains the accumulation of subretinal MPs observed in these animals.

In summary, Cx3cr1-deficient mice do not develop drusen or RPE atrophy, but provide a model of "primary" subretinal MP accumulation (when observed with age or in a nontoxic light-challenge model) that is associated with photoreceptor degeneration. In this model the subretinal inflammation develops under conditions that do not cause direct damage to photoreceptors or the RPE and the initiating cause lies within the MPs, as CX3CR1 is expressed by MPs but not by photoreceptors or RPE cells. It is particularly useful to decipher the mechanisms of MP accumulation and the effect of subretinal MPs on photoreceptor homeostasis, as the phenotype is not complicated by other reasons for photoreceptor death (such as genetic defects or toxic light conditions that directly induce cell death). The model can also be used to accentuate "secondary" models, such as a laser injury where they develop excessive inflammation and CNV, characteristics they share with CD200R$^{-/-}$ mice. Cx3cr1-deficient mice also display increased CCL2, ApoE, and IL6 expression observed in AMD (Sennlaub et al. 2013; Levy et al. 2015; Jonas et al. 2010; Anderson et al. 2001; Klaver et al. 1998; Klein et al. 2008; Seddon et al. 2005) and might therefore share important pathogenic deregulations with the human disease. Furthermore, Cx3cr1 polymorphisms have been associated with wet AMD in some studies (Combadiere et al. 2007; Anastasopoulos et al. 2012; Tuo et al. 2004) and CX3CL1 expression decreases with age in the central nervous system (Fenn et al. 2013). These results suggest that altered CX3CL1/CX3CR1 signaling might be directly involved in the athogenesis.

4.3.2.2 Defective Immunosuppressive Environment

The healthy RPE has strong immunosuppressive capacities (Wenkel and Streilein 2000) and induces MP apoptosis in vivo, evidenced by the fast clearance of MPs adoptively transferred to the subretinal space (Levy et al. 2015). In GA, MPs not only accumulate within the lesions, where the RPE is missing (Gupta et al. 2003; Combadiere et al. 2007; Penfold et al. 2001; Levy et al. 2015), but also on the RPE cells adjacent to the growing lesion (Gupta et al. 2003; Sennlaub et al. 2013; Levy et al. 2015) and on and around large drusen (Sennlaub et al. 2013; Levy et al. 2015; Killingsworth et al. 1990; Hageman et al. 2001). This accumulation might be of pathomechanistical importance as activated MPs have been shown to be able to induce RPE and photoreceptor cell death and might participate in lesion growth in GA (Sennlaub et al. 2013; Yang et al. 2011). The RPE immuno-suppressive mediators that physiologically participate in the elimination of subretinal MPs and prevent accumulation are ill defined. However, several animal models appear to present subretinal MP accumulation secondary to a defect in the RPE immunosuppressive capacities. The RPE constitutively expresses FASL (CD95L), which in part mediates its immunosuppressive characteristics (Wenkel and Streilein 2000). Indeed, FASL-defective-(*FasL^gld/gld*-mice), and FAS-defective-(*Fas^lpr/lpr*-mice) mice develop significant subretinal MP accumulation after a light challenge not strong enough to cause inflammation or damage in WT mice, similar to *Cx3cr1^GFP/GFP* mice (Levy et al. 2015). This increase in subretinal MPs is likely the result of deficient elimination, as subretinally injected MPs survived better, when FasL/FAS signaling was defective (*WT* MPs injected in *FasL^gld/gld*-mice* and *Fas^lpr/lpr*-MPs injected in WT mice compared with *WT* MPs injected in WT mice) (Levy et al. 2015). Similarly, subretinal MPs have been described to accumulate in uninjured TSP-1^−/− mice (Ng et al. 2009). One might suspect that the Crb1^rd8 mutation was present in the mice used for this report, as the typical retinal lesions are depicted in the figures (also see below). However, we recently reported the same phenotype in Crb1^rd8-free, light-challenged TSP-1^−/− mice (Levy et al. 2013). Similar to FAS/FASL deficiency, the clearance of subretinally injected MPs was impaired when TSP-1 was absent and restored by recombinant TSP-1. Interestingly, TSP-1 expression has been shown to be decreased in AMD (Uno et al. 2006) and its diminished expression might participate to create an inflammation permissive microenvironment.

Taken together, *FasL^gld/gld*-, *Fas^lpr/lpr*-, and TSP-1^−/− mice likely represent animal models in which subretinal inflammation develops due to a weakened subretinal immunosuppressive environment. There are likely other factors that play a role and future research will hopefully complete the picture.

4.3.2.3 Autoimmune Reaction

AMD shares several similarities with classical autoimmune diseases such as lymphocyte accumulation and autoantibody formation. In particular, carboxyethylpyrrole (CEP) autoantibodies have been shown to be increased in AMD plasma (Gu et al. 2010) and CEP-modified proteins accumulate in the outer retina and drusen in AMD (Crabb et al. 2002). CEP are protein modifications that form from an oxidation fragment of docosahexaenoic acid (DHA). Intriguingly, an immunization of B57BL6 mice with CEP-modified mouse serum albumin (CEPM-SA) induced a CEP-specific immune response, subretinal inflammation, and focal lesions of the RPE such as vesiculation (Hollyfield et al. 2008). Furthermore, a CEP immunization of BALB/c mice was sufficient to induce subretinal MP accumulation, followed by RPE lesions and finally photoreceptor cell loss, the main features of GA (Cruz-Guilloty et al. 2013). Interestingly, the main effector cell in this autoimmune model was shown to be CCR2+ Mo-derived iMφs, rather than a cytotoxic T-cell response (Cruz-Guilloty et al. 2013).

In summary, the above-described animal models are "primary inflammation" models, in the sense that the subretinal inflammation develops between originally healthy photoreceptors and RPE cells and under living conditions that do not directly cause RPE or photoreceptor injury. They are particularly useful to decipher mechanisms that lead to a pro-inflammatory environment in the retina and to analyze the possible downstream effect of subretinal inflammation.

4.3.3 "Secondary inflammation" AMD Models

MC activation and subretinal MP accumulation are invariably associated with models in which the integrity of either photoreceptors or the RPE is affected. Examples include models of RD, toxic light injury, and laser-induced CNV (Karlstetter et al. 2014). In models in which the blood–retinal barrier is not clearly breached, such as RD models or light-injury models, it is sometimes assumed that the subretinal MP accumulation is constituted of MCs descending from the retina. Thisconclusion might be inaccurate, as blood-borne Mo are capable to infiltrate the retina (including the subretinal space) in the absence of a detectable breach in the blood–retinal barrier. Subretinal MPs are of both MC and Mo origin in RD models (Guo et al. 2012), photo-oxidative stress (Suzuki et al. 2012), and age-related accumulation in *Cx3cr1*-deficient mice (Sennlaub et al. 2013), with no obvious barrier breach. Bone marrow-derived MPs are also not morphologically distinguishable from MCs, best evidenced by the blood-borne microglia-like cells that replace MCs after lethal whole-body irradiation (Chen et al. 2012). MP populations infiltrating the photoreceptor cell layer and sub-retinal space are likely to be constituted of a mixture of MCs and iMos in the vast majority of cases. Whether an MP is of iMos or MP origin might also be more of an academic question as activated MCs could acquire a phenotype similar to iMφs. While it is yet difficult to specifically inhibit MC activation, numerous studies have analyzed the effect of iMos recruitment, in "secondary inflammation models: depletion of circulating Mo (Sakurai et al. 2003), pharmacological inhibition of Mo recruitment (Liu et al. 2013), or genetic invalidation of the CCL2/CCR2 axis (see below), to name only a few approaches. CCL2 is a major chemokine involved in the recruitment of blood-borne iMos that express CCR2 (Geissmann et al. 2010). CCL2$^{-/-}$ and CCR2$^{-/-}$ mice have been widely used in a variety of models with secondary inflammation to evaluate the role of inflammation. CCL2 expression in the retina is physiologically low/absent but CCL2 is induced in laser injury (Yamada et al. 2007), and CNV is reduced in *Ccr2*$^{-/-}$ and *Ccl2*$^{-/-}$ mice (Liu et al. 2013; Tsutsumi et al. 2003; Luhmann et al. 2009). Similarly, CCL2 induction and CCL2/CCR2-dependent subretinal MP accumulation are important contributing factors in photoreceptor degeneration in models of photooxidative stress (Suzuki et al. 2012; Rutar et al. 2012), in the *Abca4*$^{-/-}$*Rdh8*$^{-/-}$ mouse Stargardt/AMD model

(Kohno et al. 2013), the rd10 mouse (Guo et al. 2012), in the carboxyethylpyrrole immunization-induced AMD model (Cruz-Guilloty et al. 2013), and in retinal detachment (Nakazawa et al. 2007).

Interestingly, *Ccr2*$^{-/-}$ and *Ccl2*$^{-/-}$ mice also develop subretinal MP accumulation with advanced age, which is associated with mild photoreceptor degeneration in some, but not in all laboratories (Luhmann et al. 2009; Chen et al. 2011). In our laboratory *Ccl2*$^{-/-}$ mice only developed a discreet subretinal MP accumulation, which represented only a fraction of the accumulation observed in *Cx3cr1*-deficient mice at 18 m (see supplementary data of Sennlaub et al. (Sennlaub et al. 2013)), a model of "primary inflammation (see above). In our hands, *Ccl2*$^{-/-}$*Cx3cr1*$^{-/-}$, *Ccl2*$^{-/-}$*Cx3cr1*$^{GFP/GFP}$, and *Ccr2*$^{RFP/RFP}$*Cx3cr1*$^{GFP/GFP}$ mice were partially (50 %) protected against the subretinal MP accumulation observed in *Cx3cr1*$^{-/-}$ and *Cx3cr1*$^{GFP/GFP}$ mice, but developed no associated photoreceptor degeneration, likely because the recruitment of particular neurotoxic CCR2+ iMos was prevented (see above). However, *Ccl2*$^{-/-}$*Cx3cr1*$^{-/-}$ mice in other laboratories developed focal RPE degeneration (Chen et al. 2013) or severe degeneration (Tuo et al. 2007). The latter was subsequently shown to be due to contamination with the RD 8 (Crb1^{rd8}) mutation (Mattapallil et al. 2012; Luhmann et al. 2012). In both studies it is unclear if the double knockout mice were protected against a stronger phenotype in *Cx3cr1*-deficient mice, as data from single knockout controls were not presented. Furthermore, it is not clear how CCL2 or CCR2 deletion induces subretinal MP accumulation and what the significance this observation has for AMD research as CCL2 is increased in both late forms and CCR2+ Mos accumulate in GA (see above).

The recently discovered contamination of numerous knockout mouse strains with the Crb1^{rd8} mutation (Mattapallil et al. 2012; Chang et al. 2013) has likely much affected the comprehension of AMD pathomechanisms in the last decade. Homozygous Crb1^{rd8} carriers can develop prominent focal inferior retinal lesions, photoreceptor loss, and retinal thinning. At later stages they might present vascular lesions that resemble

CNV (but are in fact derived from the retinal vasculature) and subretinal MP accumulation (Luhmann et al. 2014) similar to AMD. The phenotype is dependent on additional yet unknown genetic modifiers and the absence of macroscopic lesions does not exclude the presence of the mutation (Luhmann et al. 2014). It is likely that numerous studies using Crb1^{rd8-} contaminated knockout mice wrongly implicated genes in subretinal inflammation and the pathomechanism of AMD because of the similarity of the features caused by the Crb1^{rd8} mutation. To make matters worse, the use of Crb1$^{rd8/rd8}$ mice that are WT for the gene of interest as a control for Crb1^{rd8} carrying knockout mice cannot be accurate as long as the modifying genes are unknown.

Another important question in the role of MPs in neovascularization and degeneration is the MP polarization. MP research has classically defined two activation states in analogy to the adaptive immune system polarization Th1 and Th2: classically activated Mφs (M1) are pro-inflammatory, antimicrobial, antiangiogenic, potentially neurotoxic, and defined by the expression of mediators such as IL-1β, TNF-α, IL-6, CCL2, and iNOS. Alternatively activated Mφs (M2) are polarized toward a phenotype that is anti-inflammatory, facilitating phagocytosis, neovascularization, wound healing, and ultimately fibrosis (scarring). They are characterized by the expression of VEGF, Arginase, IL-10, and IL-1RA among others (Wynn et al. 2013; Sica and Mantovani 2012). However, these two types of polarization are only the extremes of a continuum of a multitude of activation states and all the M1/M2 defining proteins are regulated individually in the absence of a general transcriptional switch. Subretinal MPs do not generally fall into either the M1 or M2 classification, as they express M1 and M2 markers simultaneously (Horie et al. 2013; Liu et al. 2013; Camelo et al. 2012) and they are both, neurotoxic and angiogenic (see above). Therapeutically it also seems ill advised to direct subretinal MPs into a proangiogenic M2 phenotype that could encourage CNV. The identification of some M1 markers in a model and to infer that the MPs would be neurotoxic, or the identification of a selection of M2 markers to label the cells as

angiogenic does therefore not necessarily reflect a pathophysiologic reality. Furthermore, while IL-1β, TNF-α, IL-6, and VEGF likely participate in CNV the MP-derived mediators that promote degeneration are still ill defined.

Taken together, secondary subretinal inflammation occurs in all animal models where the integrity of the photoreceptors and/or RPE is altered. Its effect can be studied using a variety of pharmacological and genetic means, of which we could give here only a short overview. Although these models are clearly not specific AMD models, they allow the better understanding of the downstream effectors of subretinal inflammation that complicates a variety of pathological situations.

4.4 Conclusion

Subretinal MP accumulation is observed in early/intermediate and both late forms of AMD, but its pathogenic importance is not well defined. AMD is a complex, multifactorial disease and none of the identified genetic and environmental risk factors is sufficient to trigger the disease in humans. It is to be expected that AMD animal models mimicking a single factor (or a combination of only two risk factors) is not sufficient to produce all aspects of AMD. Interestingly, several animal models that mimic individual AMD risk factors are sufficient to induce subretinal MP accumulation but not degeneration or CNV, which might suggest that subretinal inflammation is an early feature in the pathogenesis. The use of primary and secondary subretinal inflammation models will help decipher mechanisms that lead to a pro-inflammatory environment and to analyze the possible downstream effect of subretinal inflammation. Indeed, primary inflammation models can lead to degenerative changes (Cx3cr1-deficient mice, BALB/c CEP-immunization model), and secondary inflammation is essential for CNV to develop in the laser-injury model. The use of these "AMD" models helps to understand the origin and role of subretinal inflammation in AMD and to identify drug targets to inhibit the chronic, and likely pathogenic accumulation of subretinal MPs or their neuro-toxic and angiogenic mediators.

Acknowledgments This work was supported by grants from INSERM, ANR Maladies Neurologiques et Psychiatriques (ANR-08-MNPS-003), ANR Geno 2009 (R09099DS), Labex Lifesenses, Carnot, and ERC starting Grant (ERC-2007 St.G. 210345) and HUMANIS.

Compliance with Ethical Requirements Florian Sennlaub, Emily Y. Chew and Wai T. Wong declare that they have no conflict of interest.

No human or animal studies were performed by the author for this chapter.

References

Anastasopoulos E, Kakoulidou A, Coleman AL, Sinsheimer JS, Wilson MR, Yu F et al (2012) Association of sequence variation in the CX3CR1 gene with geographic atrophy age-related macular degeneration in a Greek population. Current Eye Res 37(12):1148–1155

Anderson DH, Ozaki S, Nealon M, Neitz J, Mullins RF, Hageman GS et al (2001) Local cellular sources of apolipoprotein E in the human retina and retinal pigmented epithelium: implications for the process of drusen formation. Am J Ophthalmol 131(6):767–781

Arnson Y, Shoenfeld Y, Amital H (2010) Effects of tobacco smoke on immunity, inflammation and autoimmunity. J Autoimmunity 34(3):J258–J265

Basu SK, Ho YK, Brown MS, Bilheimer DW, Anderson RG, Goldstein JL (1982) Biochemical and genetic studies of the apoprotein E secreted by mouse macrophages and human monocytes. J Biol Chem 257(16):9788–9795

Broderick C, Hoek RM, Forrester JV, Liversidge J, Sedgwick JD, Dick AD (2002) Constitutive retinal CD200 expression regulates resident microglia and activation state of inflammatory cells during experimental autoimmune uveoretinitis. Am J Pathol 161(5):1669–1677

Burt TD, Agan BK, Marconi VC, He W, Kulkarni H, Mold JE et al (2008) Apolipoprotein (apo) E4 enhances HIV-1 cell entry in vitro, and the APOE epsilon4/epsilon4 genotype accelerates HIV disease progression. Proc Natl Acad Sci U S A 105(25):8718–8723

Calippe B, Guillonneau X, Sennlaub F (2014) Complement factor H and related proteins in age-related macular degeneration. CR Biol 337(3):178–184

Camelo S, Raoul W, Lavalette S, Calippe B, Cristofaro B, Levy O et al (2012) Delta-like 4 inhibits choroidal neovascularization despite opposing effects on vascular endothelium and macrophages. Angiogenesis 15(4):609–622

Chakravarthy U, Wong TY, Fletcher A, Piault E, Evans C, Zlateva G et al (2010) Clinical risk factors for age-related macular degeneration: a systematic review and meta-analysis. BMC Ophthalmol 10:31

Chan CC, Shen D, Zhou M, Ross RJ, Ding X, Zhang K et al (2007) Human HtrA1 in the archived eyes with age-related macular degeneration. Trans Am Ophthalmol Soc 105:92–97 (discussion 7-8)

Chang B, Hurd R, Wang J, Nishina P (2013) Survey of common eye diseases in laboratory mouse strains. Invest Ophthalmol Vis Sci 54(7):4974–4981

Chen M, Forrester JV, Xu H (2011) Dysregulation in retinal para-inflammation and age-related retinal degeneration in CCL2 or CCR2 deficient mice. PLoS ONE 6(8):e22818

Chen M, Zhao J, Luo C, Pandi SP, Penalva RG, Fitzgerald DC et al (2012) Para-inflammation-mediated retinal recruitment of bone marrow-derived myeloid cells following whole-body irradiation is CCL2 dependent. Glia 60(5):833–842

Chen M, Hombrebueno JR, Luo C, Penalva R, Zhao J, Colhoun L et al (2013) Age- and light-dependent development of localised retinal atrophy in CCL2(-/-) CX3CR1(GFP/GFP) mice. PLoS ONE 8(4):e61381

Chinnery HR, McLenachan S, Humphries T, Kezic JM, Chen X, Ruitenberg MJ et al (2011) Accumulation of murine subretinal macrophages: effects of age, pigmentation and CX(3)CR1. Neurobiol Aging.

Clark SJ, Higman VA, Mulloy B, Perkins SJ, Lea SM, Sim RB et al (2006) His-384 allotypic variant of factor H associated with age-related macular degeneration has different heparin binding properties from the non-disease-associated form. J Biol Chem 281(34):24713–24720

Clark SJ, Bishop PN, Day AJ (2010) Complement factor H and age-related macular degeneration: the role of glycosaminoglycan recognition in disease pathology. Biochem Soc Trans 38(5):1342–1348

Combadiere C, Feumi C, Raoul W, Keller N, Rodero M, Pezard A et al (2007) CX3CR1-dependent subretinal microglia cell accumulation is associated with cardinal features of age-related macular degeneration. J Clin Invest 117(10):2920–2928

Combadiere C, Raoul W, Guillonneau X, Sennlaub F (2013) Comment on "Ccl2, Cx3cr1 and Ccl2/Cx3cr1 chemokine deficiencies are not sufficient to cause age-related retinal degeneration" by Luhmann et al. (Exp. Eye Res. 2013; 107: 80.doi:10.1016). Exp Eye Res 111:134–135

Crabb JW, Miyagi M, Gu X, Shadrach K, West KA, Sakaguchi H et al (2002) Drusen proteome analysis: an approach to the etiology of age-related macular degeneration. Proc Natl Acad Sci U S A 99(23):14682–14687

Cruz-Guilloty F, Saeed AM, Echegaray JJ, Duffort S, Ballmick A, Tan Y et al (2013) Infiltration of proinflammatory m1 macrophages into the outer retina precedes damage in a mouse model of age-related macular degeneration. Int J Inflam 2013:503725

de Bont N, Netea MG, Demacker PN, Verschueren I, Kullberg BJ, van Dijk KW et al (1999) Apolipoprotein

E knock-out mice are highly susceptible to endotoxemia and Klebsiella pneumoniae infection. J Lipid Res 40(4):680–685

Fadok VA, Bratton DL, Henson PM (2001) Phagocyte receptors for apoptotic cells: recognition, uptake, and consequences. J Clin Invest 108(7):957–962

Fenn AM, Smith KM, Lovett-Racke AE, Guerau-de-Arellano M, Whitacre CC, Godbout JP (2013) Increased micro-RNA 29b in the aged brain correlates with the reduction of insulin-like growth factor-1 and fractalkine ligand. Neurobiol Aging 34(12):2748–2758

Forrester JV, Xu H, Kuffova L, Dick AD, McMenamin PG (2010) Dendritic cell physiology and function in the eye. Immunol Rev 234(1):282–304

Galea I, Bechmann I, Perry VH (2007) What is immune privilege (not)? Trends Immunol 28(1):12–18

Gautier EL, Shay T, Miller J, Greter M, Jakubzick C, Ivanov S et al (2012) Gene-expression profiles and transcriptional regulatory pathways that underlie the identity and diversity of mouse tissue macrophages. Nat Immunol 13(11):1118–1128

Gautier EL, Ivanov S, Lesnik P, Randolph GJ (2013) Local apoptosis mediates clearance of macrophages from resolving inflammation in mice. Blood 122(15):2714–2722

Geissmann F, Manz MG, Jung S, Sieweke MH, Merad M, Ley K (2010) Development of monocytes, macrophages, and dendritic cells. Science 327(5966):656–661

Grau S, Richards PJ, Kerr B, Hughes C, Caterson B, Williams AS et al (2006) The role of human HtrA1 in arthritic disease. J Biol Chem 281(10):6124–6129

Griffiths MR, Neal JW, Fontaine M, Das T, Gasque P (2009) Complement factor H, a marker of self protects against experimental autoimmune encephalomyelitis. J Immunol 182(7):4368–4377

Gu J, Pauer GJ, Yue X, Narendra U, Sturgill GM, Bena J et al (2010) Proteomic and genomic biomarkers for age-related macular degeneration. Adv Exp Med Biol 664:411–417

Guo C, Otani A, Oishi A, Kojima H, Makiyama Y, Nakagawa S et al (2012) Knockout of ccr2 alleviates photoreceptor cell death in a model of retinitis pigmentosa. Exp Eye Res 104:39–47

Gupta N, Brown KE, Milam AH (2003) Activated microglia in human retinitis pigmentosa, late-onset retinal degeneration, and age-related macular degeneration. Exp Eye Res 76(4):463–471

Haapasalo K, Jarva H, Siljander T, Tewodros W, Vuopio-Varkila J, Jokiranta TS (2008) Complement factor H allotype 402H is associated with increased C3b opsonization and phagocytosis of Streptococcus pyogenes. Mol Microbiol 70(3):583–594

Hageman GS, Luthert PJ, Victor Chong NH, Johnson LV, Anderson DH, Mullins RF (2001) An integrated hypothesis that considers drusen as biomarkers of immune-mediated processes at the RPE-Bruch's membrane interface in aging and age-related macular degeneration. Progr Retinal Eye Res 20(6):705–732

Hegner CA (1916) Retinitis exsudativa bei Lymphogranulomatosis. Klin Monatsbl Augenheil 57:27–48

Hollyfield JG, Bonilha VL, Rayborn ME, Yang X, Shadrach KG, Lu L et al (2008) Oxidative damage-induced inflammation initiates age-related macular degeneration. Nature Med 14(2):194–198

Horie S, Robbie SJ, Liu J, Wu WK, Ali RR, Bainbridge JW et al (2013) CD200R signaling inhibits pro-angiogenic gene expression by macrophages and suppresses choroidal neovascularization. Scientific Rep 3:3072

Hou Y, Lin H, Zhu L, Liu Z, Hu F, Shi J et al (2013) Lipopolysaccharide increases the incidence of collagen-induced arthritis in mice through induction of protease HTRA-1 expression. Arthritis Rheumatism 65(11):2835–2846

Huynh ML, Fadok VA, Henson PM (2002) Phosphatidylserine-dependent ingestion of apoptotic cells promotes TGF-beta1 secretion and the resolution of inflammation. J Clin Invest 109(1):41–50

Jonas JB, Tao Y, Neumaier M, Findeisen P (2010) Monocyte chemoattractant protein 1, intercellular adhesion molecule 1, and vascular cell adhesion molecule 1 in exudative age-related macular degeneration. Arch Ophthalmol 128(10):1281–1286

Karlstetter M, Scholz R, Rutar M, Wong WT, Provis JM, Langmann T (2014) Retinal microglia: just bystander or target for therapy? Progr Retinal Eye Res 45:30–57

Kerr EC, Copland DA, Dick AD, Nicholson LB (2008) The dynamics of leukocyte infiltration in experimental autoimmune uveoretinitis. Prog Retinal Eye Res 27(5):527–535

Killingsworth MC, Sarks JP, Sarks SH (1990) Macrophages related to Bruch's membrane in age-related macular degeneration. Eye (Lond) 4(Pt 4):613–621

Klaver CC, Kliffen M, van Duijn CM, Hofman A, Cruts M, Grobbee DE et al (1998) Genetic association of apolipoprotein E with age-related macular degeneration. Am J Hum Genet 63(1):200–206

Klein R, Peto T, Bird A, Vannewkirk MR (2004) The epidemiology of age-related macular degeneration. Am J Ophthalmol 137(3):486–495

Klein R, Knudtson MD, Klein BE, Wong TY, Cotch MF, Liu K et al (2008) Inflammation, complement factor h, and age-related macular degeneration: the Multiethnic Study of Atherosclerosis. Ophthalmology 115(10):1742–1749

Klein BE, Howard KP, Iyengar SK, Sivakumaran TA, Meyers KJ, Cruickshanks KJ et al (2014) Sunlight exposure, pigmentation, and incident age-related macular degeneration. Invest Ophthalmol Vis Sci 55(9):5855–5861

Kohno H, Chen Y, Kevany BM, Pearlman E, Miyagi M, Maeda T et al (2013) Photoreceptor proteins initiate microglial activation via toll-like receptor 4 in retinal degeneration mediated by all-trans-retinal. J Biol Chem 288(21):15326–15341

Kumar A, Zhao L, Fariss RN, McMenamin PG, Wong WT (2014) Vascular associations and dynamic process motility in perivascular myeloid cells of the mouse choroid: implications for function and senescent change. Invest Ophthalmol Vis Sci 55(3):1787–1796

Levy O, Raoul W, Lavalette S, Calippe B, Germain S, Guillonneau X et al (2013) Thrombospondin-1 expression controls macrophage survival in the subretinal space. Invest Ophthalmol Vis Sci 54(6):1181-

Levy O, Calippe B, Lavalette S, Hu SJ, Raoul W, Dominguez E et al (2015) Apolipoprotein E promotes subretinal mononuclear phagocyte survival and chronic inflammation in age-related macular degeneration. EMBO Mol Med (in press)

Liao S-M, Crowley M, Louie S, Delgado O, Buchanan N, Stefanidakis M et al (2013) HtrA1 regulates the subretinal infiltration of microglia cells in response to bacterial lipopolysaccharides (LPS) and aging in mice. Invest Ophthalmol Vis Sci 54(6):3666-

Liu J, Copland DA, Horie S, Wu WK, Chen M, Xu Y et al (2013) Myeloid cells expressing VEGF and arginase-1 following uptake of damaged retinal pigment epithelium suggests potential mechanism that drives the onset of choroidal angiogenesis in mice. PLoS ONE 8(8):e72935

Luhmann UF, Robbie S, Munro PM, Barker SE, Duran Y, Luong V et al. The drusen-like phenotype in aging Ccl2 knockout mice is caused by an accelerated accumulation of swollen autofluorescent subretinal macrophages. Invest Ophthalmol Vis Sci. 2009

Luhmann UF, Lange CA, Robbie S, Munro PM, Cowing JA, Armer HE et al (2012) Differential modulation of retinal degeneration by Ccl2 and Cx3cr1 chemokine signalling. PLoS ONE 7(4):e35551

Luhmann UF, Carvalho LS, Robbie SJ, Cowing JA, Duran Y, Munro PM et al (2013) Ccl2, Cx3cr1 and Ccl2/Cx3cr1 chemokine deficiencies are not sufficient to cause age-related retinal degeneration. Exp Eye Res 107:80–87

Luhmann UF, Carvalho LS, Holthaus SM, Cowing JA, Greenaway S, Chu CJ et al (2014) The severity of retinal pathology in homozygous Crb1rd8/rd8 mice is dependent on additional genetic factors. Hum Mol Genet 24(1):128–141

Lutty G, Bhutto I, Seddon J, McLeod D (2013) Mast cell degranulation in AMD choroid. Invest Ophthalmol Vis Sci 54(6):3051-

Mahley RW, Rall SC Jr (2000) Apolipoprotein E: far more than a lipid transport protein. Annu Rev Genomics Hum Genet 1:507–537

Mattapallil MJ, Wawrousek EF, Chan CC, Zhao H, Roychoudhury J, Ferguson TA et al. The rd8 mutation of the Crb1 gene is present in vendor lines of C57BL/6N mice and embryonic stem cells, and confounds ocular induced mutant phenotypes. Invest Ophthalmol Vis Sci. 2012

McKay GJ, Patterson CC, Chakravarthy U, Dasari S, Klaver CC, Vingerling JR et al (2011) Evidence of association of APOE with age-related macular degeneration: a pooled analysis of 15 studies. Hum Mutat 32(12):1407–1416

McMenamin PG, Loeffler KU (1990) Cells resembling intraventricular macrophages are present in the subretinal space of human foetal eyes. Anat Rec 227(2):245–253

Nakazawa T, Hisatomi T, Nakazawa C, Noda K, Maruyama K, She H et al (2007) Monocyte chemoattractant protein 1 mediates retinal detachment-induced photoreceptor apoptosis. Proc Natl Acad Sci U S A 104(7):2425–2430

Nathan C, Ding A (2010) Nonresolving inflammation. Cell 140(6):871–882

Ng TF, Streilein JW (2001) Light-induced migration of retinal microglia into the subretinal space. Invest Ophthalmol Vis Sci 42(13):3301–3310

Ng TF, Turpie B, Masli S (2009) Thrombospondin-1-mediated regulation of microglia activation after retinal injury. Invest Ophthalmol Vis Sci 50(11):5472–5478

Nussenblatt RB, Byrnes G, Sen HN, Yeh S, Faia L, Meyerle C et al (2010) A randomized pilot study of systemic immunosuppression in the treatment of age-related macular degeneration with *Florian Sennlaub* choroidal neovascularization. Retina 30(10):1579–1587

Penfold PL, Madigan MC, Gillies MC, Provis JM (2001) Immunological and aetiological aspects of macular degeneration. Progr Retinal Eye Res 20(3):385–414

Peri F, Nusslein-Volhard C (2008) Live imaging of neuronal degradation by microglia reveals a role for v0-ATPase a1 in phagosomal fusion in vivo. Cell 133(5):916–927

Raichlen DA, Alexander GE (2014) Exercise, APOE genotype, and the evolution of the human lifespan. Trends Neurosci 37(5):247–255

Ransohoff RM (2009) Chemokines and chemokine receptors: standing at the crossroads of immunobiology and neurobiology. Immunity 31(5):711–721

Ransohoff RM, Cardona AE (2010) The myeloid cells of the central nervous system parenchyma. Nature 468(7321):253–262

Roselaar SE, Daugherty A (1998) Apolipoprotein E-deficient mice have impaired innate immune responses to Listeria monocytogenes in vivo. J Lipid Res 39(9):1740–1743

Rosenfeld ME, Butler S, Ord VA, Lipton BA, Dyer CA, Curtiss LK et al (1993) Abundant expression of apoprotein E by macrophages in human and rabbit atherosclerotic lesions. Arterioscler Thromb 13(9):1382–1389

Rutar M, Natoli R, Provis JM (2012) Small interfering RNA-mediated suppression of Ccl2 in Muller cells attenuates microglial recruitment and photoreceptor death following retinal degeneration. J Neuroinflammation 9:221

Saederup N, Cardona AE, Croft K, Mizutani M, Cotleur AC, Tsou CL et al (2010) Selective chemokine receptor usage by central nervous system myeloid cells in CCR2-red fluorescent protein knock-in mice. PLoS ONE 5(10):e13693

Sakurai E, Anand A, Ambati BK, van Rooijen N, Ambati J (2003) Macrophage depletion inhibits experimental choroidal neovascularization. Invest Ophthalmol Vis Sci 44(8):3578–3585

Sarks SH (1976) Ageing and degeneration in the macular region: a clinico-pathological study. Br J Ophthalmol 60(5):324–341

Seddon JM, George S, Rosner B, Rifai N (2005) Progression of age-related macular degeneration: prospective assessment of C-reactive protein, interleukin 6, and other cardiovascular biomarkers. Arch Ophthalmol 123(6):774–782

Sennlaub F, Auvynet C, Calippe B, Lavalette S, Poupel L, Hu SJ et al (2013) CCR2(+) monocytes infiltrate atrophic lesions in age-related macular disease and mediate photoreceptor degeneration in experimental subretinal inflammation in Cx3cr1 deficient mice. EMBO Mol Med 5(11):1775–1793

Sica A, Mantovani A (2012) Macrophage plasticity and polarization: in vivo veritas. J Clin Invest 122(3):787–795

Sinnis P, Willnow TE, Briones MR, Herz J, Nussenzweig V (1996) Remnant lipoproteins inhibit malaria sporozoite invasion of hepatocytes. J Exp Med 184(3):945–954

Streilein JW (2003) Ocular immune privilege: therapeutic opportunities from an experiment of nature. Nature Rev Immunol 3(11):879–889

Streilein JW, Ma N, Wenkel H, Ng TF, Zamiri P (2002) Immunobiology and privilege of neuronal retina and pigment epithelium transplants. Vis Res 42(4):487–495

Suzuki M, Tsujikawa M, Itabe H, Du ZJ, Xie P, Matsumura N et al (2012) Chronic photo-oxidative stress and subsequent MCP-1 activation as causative factors for age-related macular degeneration. J Cell Sci 125(Pt 10):2407–2415

Swaroop A, Branham KE, Chen W, Abecasis G (2007) Genetic susceptibility to age-related macular degeneration: a paradigm for dissecting complex disease traits. Human Mol Genet 16(Spec No. 2):R174–R182

Tsutsumi C, Sonoda KH, Egashira K, Qiao H, Hisatomi T, Nakao S et al (2003) The critical role of ocular-infiltrating macrophages in the development of choroidal neovascularization. J Leukoc Biol 74(1):25–32

Tuo J, Smith BC, Bojanowski CM, Meleth AD, Gery I, Csaky KG et al (2004) The involvement of sequence variation and expression of CX3CR1 in the pathogenesis of age-related macular degeneration. Faseb J 18(11):1297–1299

Tuo J, Bojanowski CM, Zhou M, Shen D, Ross RJ, Rosenberg KI et al (2007) Murine ccl2/cx3cr1 deficiency results in retinal lesions mimicking human age-related macular degeneration. Invest Ophthalmol Vis Sci 48(8):3827–3836

Ufret-Vincenty RL, Aredo B, Liu X, McMahon A, Chen PW, Sun H et al (2012) Transgenic mice expressing variants of complement factor H develop AMD-like retinal findings. Invest Ophthalmol Vis Sci 51(11):5878–5887

Uno K, Bhutto IA, McLeod DS, Merges C, Lutty GA (2006) Impaired expression of thrombospondin-1 in eyes with age related macular degeneration. Br J Ophthalmol 90(1):48–54

Wang Y, Wang VM, Chan CC (2011) The role of anti-inflammatory agents in age-related macular degeneration (AMD) treatment. Eye (Lond) 25(2):127–139

Wei L, Liu B, Tuo J, Shen D, Chen P, Li Z et al (2012) Hypomethylation of the IL17RC promoter associates with age-related macular degeneration. Cell Rep 2(5):1151–1158

Wenkel H, Streilein JW (2000) Evidence that retinal pigment epithelium functions as an immune-privileged tissue. Invest Ophthalmol Vis Sci 41(11):3467–3473

Wynn TA, Chawla A, Pollard JW (2013) Macrophage biology in development, homeostasis and disease. Nature 496(7446):445–455

Xu H, Chen M, Manivannan A, Lois N, Forrester JV (2008) Age-dependent accumulation of lipofuscin in perivascular and subretinal microglia in experimental mice. Aging Cell 7(1):58–68

Yamada K, Sakurai E, Itaya M, Yamasaki S, Ogura Y (2007) Inhibition of laser-induced choroidal neovascularization by atorvastatin by downregulation of monocyte chemotactic protein-1 synthesis in mice. Invest Ophthalmol Vis Sci 48(4):1839–1843

Yang Z, Camp NJ, Sun H, Tong Z, Gibbs D, Cameron DJ et al (2006) A variant of the HTRA1 gene increases susceptibility to age-related macular degeneration. Science 314(5801):992–993

Yang D, Elner SG, Chen X, Field MG, Petty HR, Elner VM (2011) MCP-1-activated monocytes induce apoptosis in human retinal pigment epithelium. Invest Ophthalmol Vis Sci 52(8):6026–6034

Yu Y, Reynolds R, Rosner B, Daly MJ, Seddon JM (2012) Prospective assessment of genetic effects on progression to different stages of age-related macular degeneration using multistate Markov models. Invest Ophthalmol Vis Sci 53(3):1548–1556

Yu CE, Cudaback E, Foraker J, Thomson Z, Leong L, Lutz F et al (2013) Epigenetic signature and enhancer activity of the human APOE gene. Hum Mol Genet 24(1):128–141

Zieger M, Ahnelt PK, Uhrin P (2014) CX3CL1 (Fractalkine) Protein expression in normal and degenerating mouse retina: in vivo studies. PLoS ONE 9(9):e106562

Commentary

Emily Y. Chew

e-mail: echew@nei.nih.gov

Division of Epidemiology and Clinical Applications,

National Eye Institute,

National Institutes of Health,

Bethesda, MD, USA

Age-related macular degeneration (AMD), a leading cause of blindness in the developed world, has two late forms of disease, neovascular AMD and geographic atrophy, which lead to the loss of the photoreceptor and the retinal pigment epithelium. Although there are remarkable treatments for neovascular AMD, the underlying process of atrophy of AMD continues to challenge us. There is no proven therapy for geographic atrophy and it continues to progress often in eyes treated successfully for neovascular AMD. Although its pathogenesis is unknown, the wellknown risk factors associated with the development of AMD includes increasing age, genetic variants, and environmental factors.

Developing animal models to study human AMD is essential to our understanding of its pathogenesis and testing of potential therapies for this disease. For example, although we have now demonstrated that there are more than 34 genetic loci associated with AMD, we do not have an understanding of the functional impact of these genetic variants on protein function, etc. Animal models would be helpful in elucidating the mechanisms of disease. The genetic loci point to potential inflammation, lipid metabolism, angiogenesis, ECM remodeling, and mitochondrial effect. Using animal models, more specifically mouse models, these processes can be evaluated as to whether these pathological mechanisms are important in AMD. Despite the lack of the macula in the mouse model and the inability to replicate all the characteristics of AMD in the mouse model, such as retinal pigment epithelial atrophy, the animal models will help us understand the basic processes involved in the pathobiology of AMD. The relative ease of genetic manipulation, relatively cost-effectiveness with a somewhat accelerated time course, makes these animal models appealing. Incorporating the risk factors such as age, environmental stressors along with the genetic manipulations, researchers may achieve the goal of targeting drug development for the treatment of geographic atrophy associated with AMD.

Commentary

Wai T. Wong
e-mail: wongw@nei.nih.gov
Unit on Neuron-Glia Interactions in Retinal Disease,
National Eye Institute,
National Institutes of Health,
Bethesda, MD, USA

As many retinal pathologies have chronic retinal inflammation as a shared feature, cellular changes in the physiology and distribution of immune cells within the retina constitute a significant disease-related phenotype relevant to the evaluation and interpretation of animal models of disease. In AMD, the accumulation of innate immune cells in the forms of microglia, Mo, macrophages, and/or DCs in the subretinal space have received much scrutiny in AMD-related animal models. Although this feature may possibly be an epiphenomenon linked to other AMD phenotypes, or even a protective response, subretinal inflammation has been predominantly interpreted as a dysregulation of the outer retinal immune environment that contributes to the progression of photoreceptor/RPE degeneration and CNV. If the latter possibility is in fact true, understanding of the mechanisms driving and sustaining subretinal inflammation and its consequences in the retina will be very insightful in elucidating AMD pathogenesis, and may even give rise to new therapeutic strategies for AMD related to this phenomenon.

In this chapter, the AMD-related animal models featuring subretinal inflammation are outlined and discussed. The author has highlighted some dichotomies to help guide the categorization of these models: one contrast is drawn between "first-degree" models, where perturbations in AMD-related causative factors (e.g., genetic risk factors) result in AMD-related phenotypes, versus "downstream" models where AMD-related phenotype is produced by factors not directly linked to AMD pathogenesis. A second dichotomy is drawn between "primary inflammation" models, where subretinal inflammation occurs as a result of altered immune regulation involving the implicated immune cells, or induced by autoimmune responses, versus "secondary inflammation" models, where subretinal inflammation is subsequent to an injury- or degeneration-related trigger. Because these models involving subretinal inflammation are varied in phenotype and have diverse etiologies, considering where the model lies with respect to these dichotomies can be helpful in judging how relevant or insightful the derived observations are to AMD in humans. As a general guideline, the author proposes "first-degree" and "primary inflammation" models as having greater relatedness and explanatory power over their counterparts. However, the fidelity and completeness of the AMD-associated phenotypes capitulated in the model may also be an important consideration.

These perspectives and considerations outlined in this chapter are helpful in putting the phenomenon of subretinal inflammation in the broad context of AMD pathogenesis. For example, there has been considerable discussion in the field regarding the interpretation and use of the $CX3CR1^{-/-}/CCl2^{-/-}$ Crb^{rd8} model; in one perspective, the model may be considered less useful on the basis that it is a "downstream" model involving "secondary inflammation" model. In a counter perspective, the model may be argued as recapitulating many aspects of AMD-related pathology and thus has value and relevance. Despite these disagreements, it is likely that further examination of subretinal inflammation, and its causes, regulators, and consequences, will be insightful and productive in the elucidation of AMD pathogenesis.

Animal Models of Diabetic Retinopathy

5

Mei Chen and Alan Stitt

5.1 Introduction and Clinical Context

Diabetic retinopathy (DR) is the most commonly occurring microvascular complication of type-1 and type-2 diabetes (T1D; T2D; Antonetti et al. 2012). The occurrence and progression of DR is linked to hyperglycaemia, hypertension, renal disease and dyslipidaemia in association with diabetes duration (Stitt et al. 2013). In particular, the importance of poor glycaemic control has been proven by the Diabetes Control and Complications Trial (DCCT) for T1D (DCCT 1993) and by the UK Prospective Diabetes Study (UKPDS) for T2D (UKPDS1998).

Diabetes impacts on all cells of the retina, although, most attention has been focused on retinal microvascular pathology and DR classification has been based on these lesions. The standard Early Treatment of Diabetic Retinopathy Study (ETDRS) scale (ETDRS 1991) is based on the number of photographically detectable microvascular lesions in the retina. Using ETDRS, clinical DR has been classified into two stages based on the level of microvascular degeneration and its related ischaemic damage to the retina: the early, non-proliferative DR (NPDR) and the

advanced, proliferative DR (PDR). The NPDR can be further classified into (i) mild NPDR (presence of microaneurysms in the retina); (ii) moderate NPDR (presence of small haemorrhages and 'cotton wool' spots); and (iii) severe NPDR (haemorrhages, hard exudates, patches of ischaemia; Stitt et al. 2013). The progression of NPDR from mild to severe is related to progressive retinal microvascular damage, including vascular endothelial cell and pericyte loss, vasoobliteration, occlusion of capillaries, thickening of vascular basement membrane (BM), and abnormal permeability of the blood retina barrier (BRB). Figure 5.1a shows a colour fundus photograph of NPDR with significant macular oedema. Microaneurysms can be identified in Fig. 5.1b and show a predominantly peri-arterial distribution.

If left untreated, NPDR can develop to PDR which is the presence of new blood vessels or neovascular membranes in the pre-retinal space. PDR typically occurs in approximately 25 % of patients with T1D and in about 15 % of patients with T2D with > 25 years disease duration (Klein et al. 1984). Progressive retinal microvascular damage leads to severe ischaemia and the release of vasoproliferative factors from retinal cells, which subsequently stimulate new blood vessels to grow in the retina in order to improve oxygenation. PDR is closely associated with retinal ischaemia and neovascularisation is typically observed in areas adjacent to extensive vascular non-perfusion. These new vessels are fragile causing haemorrhages in the retina and vitreous.

A. Stitt (✉) · M. Chen
Centre for Experimental Medicine, Queen's University Belfast, Northern Ireland, UK
e-mail: a.stitt@qub.ac.uk

M. Chen
e-mail: m.chen@qub.ac.uk

© Springer International Publishing Switzerland 2016
C.-C. Chan (ed.), *Animal Models of Ophthalmic Diseases,* Essentials in Ophthalmology,

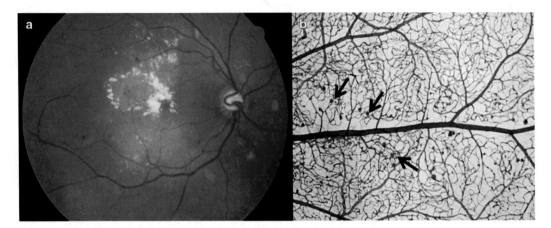

Fig. 5.1 (a) Fundus photograph of the right eye of a patient with NPDR and central involvement clinically significant macular oedema. Marked exudation is observed around the fovealcentre. (b) Trypsin digest retinal specimen of a 65-year-old type 2 diabetic patient with NPDR stained with PAS-haematoxylin. A centrally placed radial artery is flanked above and below by retinal veins and the intervening capillary beds. Numerous microaneurysms stain strongly with PAS and show a predominantly peri-arterial distribution *(arrows)*

In addition, fibrovascular membranes may form around the neovascular site and their contraction may cause retinal tears and tractional detachment, resulting in severe vision loss.

When diabetes impacts the macula, it is termed diabetic maculopathy. In this area of central vision, excessive vasopermeability and excess fluid in the retina is called diabetic macular oedema (DMO). DMO typically occurs during the late stages of disease and is the most frequent cause of blindness in diabetic patients (Joussen et al. 2007). The leakage of plasma proteins from the damaged BRB is the primary cause of DMO, which results in the accumulation of fluids within the retina, in particular within and/or around the macular area. DMO can readily be detected by optical coherence tomography (OCT), and is characterised by increased retinal thickness and/or the presence of fluid-filled cysts in the macular region (Fig. 5.2a and 5.2b). DMO is more common than PDR, although, both endpoints can occur together.

While loss of retinal perfusion is a hallmark of DR, neural function is also compromised during this disease, and it has been suggested that this could occur prior to overt vessel pathology (reviewed by Antonetti et al. (2012)). Electrophysi-ological studies of patients with diabetes display loss of colour vision (Roy et al. 1986) and contrast sensitivity (Sokol et al. 1985) and they often show abnormalities in the electroretinogram (ERG; Tzekov and Arden 1999). DR may therefore be more accurately conceived as a disease of the neuro-vascular unit, resulting in dysfunction and eventual death of several vascular cells, neurons and glia (Antonetti et al. 2012).

Prevention of DR in patients is reliant on metabolic control relating to maintenance of tight glycaemic control, blood pressure control and correction of dyslipidaemia (Stitt et al. 2013). When patients progress to the late stages of DR, there are several treatments such as photocoagulation, intravitreal delivery of vascular endothelial growth factor (VEGF) blocking agents and intravitreal corticosteroids, all of which have proven value in reducing DMO. Laser photocoagulation and vitreoretinal surgery are the main treatments for PDR. Figure 5.2c and 5.2d shows a typical colour fundus photograph (Fig. 5.2c) and a fundus fluorescein angiography (Fig. 5.2d) of the eye with PDR treated with panretinal photocoagulation. Unfortunately, current approaches to DMO and PDR are, by definition, at the late stage and carry significant side effects. Further-

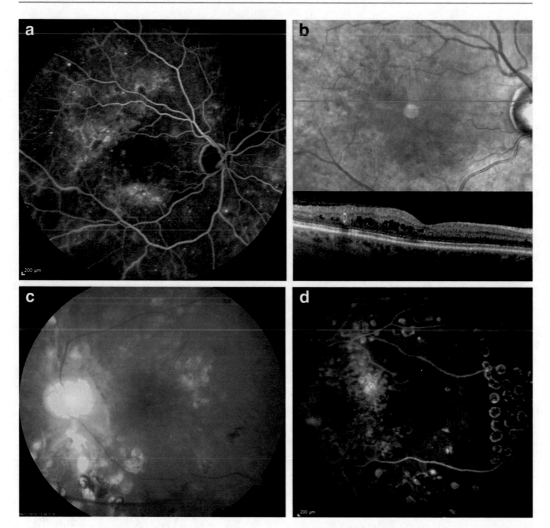

Fig. 5.2 Fundus fluorescein angiography (FFA) (**a**) and horizontal optical coherence tomography (OCT) (**b**), with the corresponding fundus image demonstrating the location from which the OCT section had been taken, obtained from the RE of a patient with diabetic macular oedema. On FFA, two areas of hyperfluorescence, located superotemporally and inferior to the fovea, were detected. On OCT, intraretinal oedema extending to fixation was evident. Colour fundus photograph (**c**) and fundus fluorescein angiography (FFA) (**d**) of the *left* eye of a patient with proliferative diabetic retinopathy treated with extensive panretinal photocoagulation. Despite the latter treatment recurrent vitreous haemorrhages occurred. Fluorescein angiography was undertaken once the last vitreous bleeding had cleared and disclosed an area of retinal ischaemia which had been left untreated temporal to the fovea. This area seemed to support the persistence of a frond of active neovessels located in the nasal retina, whose presence was also demonstrated by FFA (not shown on the images)

more, they fail to address the underlying cause of ischaemia and/or inflammatory pathology which drives these endpoints. It is clear that there are deficits in the current standards of care for DR and that there is a need for improved therapies, especially for the ones that show efficacy at the early stages of disease. The use of animals that model key aspects of the structural, functional and metabolic features of DR plays a critical role in understanding the disease and development of new therapeutic approaches.

5.2 The Need for Animal Models to Understand the Pathophysiology of DR

Retina-specific and systemic molecular defects that occur during diabetes make DR pathogenesis multifactorial and highly complex and many interacting pathogenic pathways have been proposed to play causative roles, as reviewed by Antonetti et al. (2012) and Brownlee M (2001). This complexity and the decades-long progression of DR present a serious challenge for developing animal models of the clinical disease. For example, the various animal species used have differing retinal anatomy, metabolism, diets, lifespans, etc. and while they may be able to reproduce some aspects of human diabetes, no single model currently exists that faithfully reproduces all aspects of human DR. In particular, it is important to note that many key lesions that are regarded as typical for background DR in patients, including microaneurysms, exudates, haemorrhages and cotton wool spots seldom, if ever, occur in short-term diabetic rodent models and only appear in dog, pig or monkey models with longer durations of diabetes (Lai and Lo 2013). Moreover, there are no mainstream T1D or T2D models that reproduce DMO or PDR endpoints. Therefore the value of various animal models currently available needs to be carefully evaluated to avoid injudicious extrapolation to clinical scenarios. In this section, we will discuss the merits and shortcomings of common animal models.

5.3 Rodent Models of DR

5.3.1 Chemically Induced Diabetes in Mice and Rats

The capacity for breeding and housing large numbers of animals in a relatively small facility has made rats and mice the most common model for DR research. While they bring cost, uniformity and transgenic technology advantages, rodents have obvious disadvantages such as metabolism, diet and lifestyle that are very different from humans. Mice and rats species used in laboratories

evolved as nocturnal animals and, as such, their eyes are adapted for night vision with a large lens and a rod-dominant retina, with few cones and no macula. However, mice and rats do have an intraretinal vasculature and a neuroglial structural anatomy that is similar to the humans and this has made rodent models of DR the predominant species for research.

Streptozotocin (STZ) or alloxan are often used as chemicals to destroy pancreatic β-cells and thereby induce T1D in rodents. STZ is used most commonly to induce diabetes in rats of various strains and they have been the mainstay for many of the seminal studies demonstrating retinal lesions after diabetes durations of 1–12 months (Robinson et al. 2012). In both mice and rats, diabetes leads to dysfunction or failure (or partial failure) of various organs, including the retina, kidneys, nerves and heart. Depending on the strain used and the duration of hyperglycaemia, these diabetic models reproduce some early lesions of DR, including thickening of the vascular BM, increased vascular permeability, loss of retinal pericytes and capillary closure (Robinson et al. 2012). Nonvascular changes, including biochemical changes, neuronal and glial changes are often detected and might contribute to the vascular pathology observed in these models. Figure 5.3 shows some representative DR features observed in STZ-induced rodent models.

Rodent models of DR have permitted the precise regulation of elements in the diabetic milieu, intervention with experimental drugs that modulate gene regulation and detailed cellular and molecular analysis of the retina. For example, studies in animal models of DR have strengthened our appreciation of pro-inflammatory pathways and their connectivity to lesions such as BRB breakdown. Using detailed experimental approaches in animals, especially rodents, it has been established that indicators of inflammation, including leukostasis, increased expression of adhesion molecules in retinal vascular endothelial cells and leukocytes, altered vascular permeability, and increased production of prostaglandins, nitric oxide, cytokines and other inflammatory mediators occurs in the retina after diabetes durations of 1–6 months (Tang and Kern

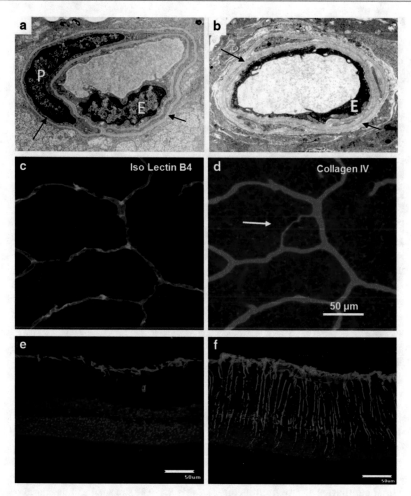

Fig. 5.3 Transmission electron microscopy (TEM) shows the thickened basement membrane *(arrows)* and loss of viable endothelial cell (E) in a diabetic rat (**b**). A typical normal retinal capillary (**a**) showing endothelial cell (E) and pericyte (P) with the cohesive basement membrane *(arrows)*. Acellular capillary can be detected via the immunofluorescent staining technique. Acellular capillary *(arrow* in **d**) loses endothelial cells (Isolectin B4, **c**) but persists with the basement membrane (Collagen IV, **d**). Müller cell activation can be detected in diabetic rodents. Expression of GFAP is limited at the end feet of Müller cells in the normal retina (**e**); however, GFAP can be detected across the whole retina in a diabetic rodent (**f**). *Red*: GFAP; *blue*: DAPI

2011). Microglia are the resident macrophage in the retina and they play an important role in maintaining homeostasis in the neuropile. These cells become activated in diabetes and contribute to the pro-inflammatory state. Microglia become altered in DR, with the cell morphology changing from a resting, surveillance state with long thin dendrites emanating from a small cell body (Fig. 5.4a and 5.4b) to an activated, amoeboid state with thick dendrites and an enlarged cell body (Fig. 5.4c and 5.4d).

It is important to note that the retinal biochemistry and histopathological features differ between rat and mice and even between different strains of the same species. For example, pigmented Brown Norway rats develop sustained vascular leakage during the experimental period of 16 weeks of diabetes, whereas the albino Sprague–Dawley rats typically show retinal hyperpermeability from day 3 to day 10 after the onset of diabetes (Zhang et al. 2005). Likewise, Lewis rats seem to demonstrate the most acceler-

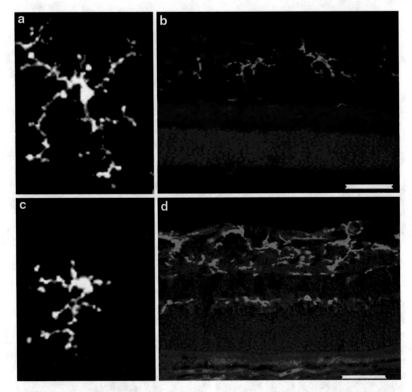

Fig. 5.4 Microglial activation in a diabetic retina. Microglia morphology changes from a resting state with long dendrites and small cell body (**a**) to an activated, amoeboid state with shorter dendrites (**c**). Increased immunoreactivity is also observed in retinal cross-section of a diabetic rat (**d**) compared with a normal retina (**b**). *Green*: microglia; *blue*: diamidino-2-phenylindole (DAPI)

ated loss of retinal capillaries after 8 months diabetes as compared to Sprague–Dawley rats that show fewer lesions at the same time points (Kern et al. 2010).

In recent years, mice have been commonly used in preference to rats, since they have reduced costs and provide greater availability of molecular tools and offer scope from transgenic technology. Although transgenic technology is not confined to mice (and transgenic diabetic rats have been used to great effect (Berner et al. 2012)), these murine models have provided a significant value in dissection of molecular mechanisms. The structural and functional lesions in diabetic mice are summarised in Table 5.1. In comparison to rats, mice are comparatively resistant to diabetes induction using STZ and considerably higher doses of STZ are required. While some studies have used a single high dose of STZ

Table 5.1 The structural and functional lesions in diabetic mice

Structural lesions	Functional lesions
Vasculopathy	
Capillary basement membrane thickening	Leukostasis
Capillary degeneration, endothelial cell death and pericyte drop-off	BRB dysregulation of inner retina
Neuropathy	
Retinal degeneration	Abnormal retinal electrophysiology (reduced ERG a-, b-wave amplitude, prolonged oscillatory potentials implicit time
Neuronal cell damage, loss of synaptic connectivity	
Ganglion cell apoptosis	

in mice (Cox et al. 2003), this may produce less desired, non-beta cell effects. Commonly, five injections of low-dose STZ given over successive days has been the standard protocol for induction of diabetes in mice (McVicar et al. 2011).

5.3.2 Spontaneous Diabetic Rodents as Models of DR

Mice and rats can develop diabetes spontaneously either due to mutations in certain insulin-related genes or due to a susceptibility to obesity and insulin resistance. These animals can provide the advantage that their T1D or T2D diabetes is naturally occurring and is not induced by chemicals that could have possible non-diabetes-related effects in several organs, especially when high doses are used. The disadvantages of these models often relate to their increased cost because they are not widely bred and, in some cases, diabetes only occurs in homozygous or in a single sex which results in the need to breed much higher numbers of animals. Rat strains that develop spontaneous T1D or T2D include Zucker diabetic fatty rats, WBN/Kob rats, Otsula Long-Evans Tokushima fatty rats, Goto-Kakizaki (GK) rats and diabetic Torii rats. A number of mouse strains also develop diabetes spontaneously; these include nonobese diabetic mice (NOD), db/db mice and OB/OB mice which develop T1D and T2D, respectively. Retinal biochemistry and histopathological features such as vascular lesions (Abari et al. 2013) and neural retina thinning (Bogdanov et al. 2014) have been reported in these DR models.

Of particular note, the Ins2^{Akita} mouse is a model of spontaneous T1D, which has been shown to develop a range of vascular, neural and glial lesions that are comparable to what has been shown in STZ diabetic mice (Barber et al. 2005; Gastinger et al. 2006). Although one study reported no retinal thinning or vascular changes in Ins2^{Akita} mice up to 6 months of age (McLenachan et al. 2013), a more recent study has shown a progressive thinning of the retina from 3 months onwards and the loss of synaptic connectivity, particularly in second-order neurons in these mice (Hombrebueno et al. 2014). The Ins2^{Akita} mice is widely used in experimental studies of diabetes complications, including DR.

5.3.3 Diet-Induced DR in Rodents

An enriched diet can also induce diabetes. For example, high-fat fed mice develop early T2D, with impaired glucose intolerance and insulin resistance (Winzell and Ahren 2004). Rodents fed with a galactose-rich diet for a prolonged period of time develop a retinopathy that resembles the early stages of DR, including capillary dropout and pericyte loss (Kern and Engerman 1994).The galactose-fed model was initially used for investigating the role of polyol pathway in the pathogenesis of DR (Robison et al. 1983). Studies suggested that aldose reductase inhibitors (ARIs) reduced the incidence and severity of diabetic retinal lesions occurring in these galactose-fed animals (Robison et al. 1983).

An interesting diet-induced rodent model of DR has been reported in the sand rat (*Psammomysobesus*). These animals are active in daytime and have cone-rich retinas unlike nocturnal laboratory rats and mouse strains. When sand rats are maintained in captivity for 4–7 months and fed standard chow, ~60 % develop T2D-like symptoms and show facets of DR such as BRB dysfunction, loss of retinal neurons and capillary degeneration (Saidi et al. 2011). There is also a suggestion that these animals may show some degree of pre-retinal neovascularisation (Saidi et al. 2011), although, how frequently this occurs and the underpinning pathology remains ill-defined.

5.4 Dog Models of DR

Next to rodents, dogs have been used most extensively as models of DR and although there are obvious ethical constraints and significant cost issues, this species is an excellent model of the early-stage clinical disease. Most studies have used chemical induction of diabetes (Anderson et al. 1993) or feeding with galactose (Kador et al. 1994), although the same caveats exist when comparing hyperglycaemia with galactosaemia (Engerman et al. 1990). The main advantage of dogs over mice and rats is the duration of diabetes that can be achieved due to longer lifespan of this species and, typically, 5 years hyper-

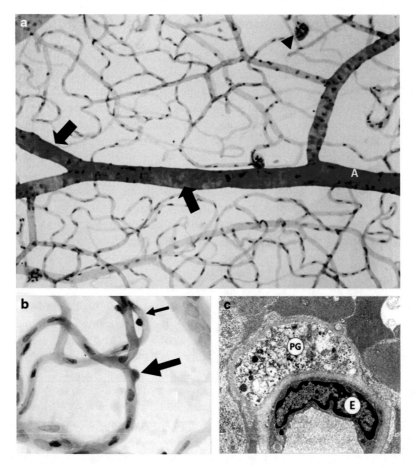

Fig. 5.5 a Retinal vasculature from a 5-year-old diabetic dog using trypsin digest showing microaneurysm *(arrow head)* and widespread loss of arteriolar smooth muscle (SM) cells *(arrows)*. **b** Trypsin digest of a 4-year-old diabetic dog showing a pericyte ghost stained *red* by the PAS technique *(large arrow)*. A viable pericyte staining strongly with haematoxylin is also apparent on the capillary wall *(small arrow)*. **c** Electron micrograph of a 4-year-old diabetic dog shows a pericyte ghost (PG) represented as a pocket of vesicular debris within the basement membrane (BM). The endothelial cell (E) remains viable. Lumen (L) (With permission of Springer Science+Business Media)

glycaemia or galactosaemia produces lesions that are much closer to clinical NPDR than what is possible with rodent models (Howell et al. 2013). With daily insulin delivery, the degree of hyperglycaemia in diabetic dogs can be maintained to much lower levels, which provides a more clinically relevant disease state than can be achieved with insulin-treated diabetic rodents (Anderson et al. 1993; Gardiner et al. 1994). Notably, dogs which have been diabetic for 5 years show microaneurysms and loss of arteriolar smooth muscle (Gardiner et al. 1994, Fig. 5.5), lesions that are rarely, if ever, observed in diabetic rodents.

Dogs fed with a galactose-rich diet for a prolonged period of time develop a retinopathy that resembles the early stages of NPDR (Engerman et al. 1990). Galactosaemia leading to DR-like lesions in dogs has been more extensively reported than induced diabetes (Engerman and Kern 1984; Kern and Engerman 1996; Robison et al. 1983) and this model shows retinal dot and blot haemorrhages, pericyte death, capillary loss, microaneurysms starting after approximately 3 years and becoming more confluent after 6 years feeding (Engerman and Kern 1984; Kern and Engerman 1996; Robison et al. 1983). Other retinal vascular lesions, including intraretinal microvessels, arterial occlusion and formation of arteriovenous shunts, have been reported (Kador et al. 2002). Interestingly, retinal neovascular lesions have been described after prolonged galactose feeding (for up to 84 months; Kador et al. 2002; Cusick et al. 2003).

Since the profile of DR in dogs is much closer to the clinical scenario, this species has been used to assess drug efficacy (Gardiner et al. 2003b; Kador et al. 1990; Kowluru et al. 2000). Moreover, diabetes induction in dogs has been used as strong evidence for the glycaemic memory phenomenon observed in patients (DCCT 2014). Engerman et al. studied four experimental groups of dogs including diabetics, non-diabetics, poor control diabetics, good control diabetics and diabetic dogs that were switched from poor to good glycaemic treatment (Engerman and Kern 1987). Diabetic dogs that were switched to good glycaemic control using insulin injections after 2.5 years of poor glycaemic control, continued to develop retinopathy similar to the poor glycaemic control group (Engerman and Kern 1987). This dog study utilised the long-term advantages of this experimental model to demonstrate that early intensive glycaemic control is capable of delaying the development of DR while an early period of poor glycaemic control carries increased risk.

5.5 Other Large Animals of DR

Pig retina has a cone-rich region known as the *area centralis* which is somewhat analogous to the primate macula. This combined with the large eye-size for imaging and surgery provides a significant advantage as a model of human DR. While this model has not been extensively reported in comparison to rodents (and even dogs), chemical diabetes induction in pigs is achievable (King et al. 2011) and has often been combined with high-fat diet (Hainsworth et al. 2002). Following 4 months diabetes, pig retina shows pericyte degeneration, vascular BM thickening and a compromise of the BRB (Lee et al. 2010).

Although not a commonly used model of DR, diabetes induction in cats (T1D) also produces retinal lesions such as capillary BM thickening, vessel tortuosity, BRB dysfunction, capillary degeneration and microaneurysms (Mansour et al. 1990; Hatchell et al. 1995).

Non-human primate models of diabetes provide the ultimate model of clinical DR since these animals are anatomically closest to humans and the retina has a macula. There have been several, high-value studies examining retinopathy following T1D and T2D in rhesus monkeys. The Lutty and Tso groups have made a significant contribution to this area and they have reported DR lesions in spontaneously T2D monkeys (Johnson et al. 2005; Kim et al. 2004) and in monkeys with STZ-induced T1D (Tso et al. 1988; Buchi et al. 1996). These studies have shown, perhaps unsurprisingly, that in primates, DR follows a more clinically linked profile, with variation between subjects but some severe retinopathy occurring especially in the T2D models with up to 12 years of disease. Neural retina abnormalities were present, including ERG defects (Kim et al. 2004) and the lesions included extensive retinal ischaemia with the presence of neural retina infarctions (cotton-wool spots), extensive microaneurysms, capillary degeneration, large vessel occlusions, BRB damage, intraretinal microvascular abnormalities (IRMA) and atrophy of the macula (Johnson et al. 2005). Interestingly, despite of extensive ischaemia, no advanced PDR was apparent (Johnson et al. 2005).

These larger animals have many advantages as models of DR, although the cost disadvantages and inability to use large experimental groups are obvious. The use of non-human primates as models of human disease is also challenging from an ethical point of view and limits the use of these models in DR research.

5.6 Non-mammalian Models of DR

The high cost and ethical constraints of mammalian models of DR have led to the evaluation of the zebrafish (*Daniorerio*) as a viable model for some aspects of human disease. These fish have been extensively characterised from the embryological, anatomical and genetic point of view, and combined with their rapid reproductive timescale and improved imaging technology, they provide opportunities for high-throughput in vivo modelling. In the context of DR research, zebrafish have had diabetes induced by direct STZ injection with hyperglycaemia achieved (Olsen et al. 2010). Interestingly, with this diabetes-inducted model, zebrafish show impaired limb regeneration when compared with non-diabetic counterparts (Olsen et al. 2010) and there is enticing evidence that the glycaemic memory phenomenon also occurs in these diabetic fish (Intine et al. 2013; Sarras et al. 2013; Olsen et al. 2012).

Some groups have also exposed zebrafish to 'hyperglycaemic-like' conditions by adding high glucose to their water. Zebrafish exposed to high glucose for 28 days resulted in neural retina degenerative changes when compared with normal glucose controls (Gleeson et al. 2007). Furthermore, oscillating high glucose exposure for 30 days caused the zebrafish to lose visual function linked to cone photoreceptor degeneration and a degree of BM thickening and BRB damage (Alvarez et al. 2010), although it should be appreciated that these fish have a pre-retinal hyaloid vasculature.

5.7 Models of PDR

The overwhelming majority of DR models reproduce key aspects of NPDR and there are no robust systems to produce a pre-retinal neovascularisation that is comparable to human PDR. Thus, scientists have been using retinal neovascularisation in non-diabetic animal models or utilised transgenic approaches to over express VEGF in diabetic mice.

5.7.1 VEGF Overexpression

Developed by the Rakoczy group, the Kimba mouse is a transgenic model of sub-retinal neovascularisation (van Eeden et al. 2006). This line of mice was generated by overexpressing the human *VEGF165* gene in the photoreceptors under control of rhodopsin promoter. These mice develop spontaneously retinal vascular leakage, microaneurysms, capillary dropout and intraretinal neovascularisation as a model for age-related macular degeneration (van Eeden et al. 2006; Ali Rahman et al. 2011). This model has derived a new mouse which is more relevant to DR created by crossing Kimba mice with the Ins2[Akita] mice (Rakoczy et al. 2010). The so-called Akimba mouse inherits the key features from the parental strains: overexpression of VEGF in the photoreceptor and spontaneous T1D. Vascular abnormalities, including microaneurysms, vascular leakage, venous beading and capillary dropout, have been observed in this model (Wisniewska-Kruk et al. 2014) which is more linked to the overexpression of human VEGF in the photoreceptors rather than hyperglycaemia. While the Akimba mouse has value, it does not develop ischaemia-induced pre-retinal neovascularisation, such as the occurrence in PDR, and hence may not be ideal for studying the aetiology of this late-stage pathology in DR.

5.7.2 Oxygen-Induced Retinopathy

Oxygen-induced retinopathy (OIR) was originally developed by the Smith laboratory to study retinopathy of prematurity (ROP) in newborns (Smith et al. 1994). In this model, postnatal day 7 animals (P7) are exposed to elevated partial pressure of oxygen (~ 75 % oxygen) for 5 days after which normal retinal vasculature development ceases and existing intraretinal capillaries in the central retina degenerate due to a fall in VEGF (Simpson et al. 1999). Figure 5.6a demonstrates an ischaemia mouse retina after exposing to 75 % oxygen for 5 days. The animals are then removed from the high-oxygen environment to normal atmospheric conditions (21 % oxygen) after which the retina suffers severe ischaemia due to insufficient blood and oxygen supply and pre-retinal neovascularisation ensues. Retinal hypoxia is a feature of ischaemic retinopathies such as PDR, and the OIR model produces this in a highly delineated manner in the central retina (Fig. 5.6b). Oxygen deprivation of the retinal glia and neurons drives pro-angiogenic growth factor ex-

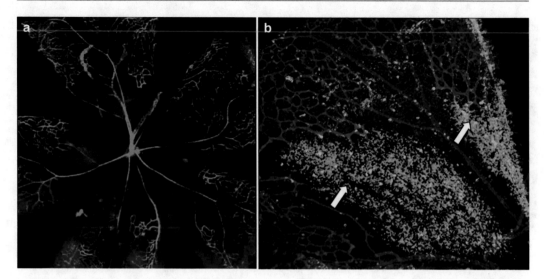

Fig. 5.6 Retinal flatmount immuno-fluorescent staining demonstrates the ischaemia/hypoxia retina from a 12-day-old mouse with oxygen-induced retinopathy. **a** Blood vessels are detected using iso-lectin B4 *(green)*. **b** Hypoxia-sensitive drug pimonidazole (*green* fluorescence) are deposited in the avascular areas. The blood vessels are *red* stained with Isolectin B4 (With permission of Springer Science+Business Media)

pression in OIR and the neovascular membranes that ensue are similar to those occurring in PDR (Connor et al. 2009). The pathology is consistent, reproducible and readily quantified (Connor et al. 2009) and the model has been extensively used for assessment of vasogenic agents (Gardiner et al. 2003a; Gebarowska et al. 2002; McVicar et al. 2011) and cell therapies (Medina et al. 2010; Medina et al. 2011). In addition, BRB breakdown, glial and neuronal damage, and reduced retinal thickness have been observed in the OIR model (Vessey et al. 2011). The OIR model (and other neonatal rat-based variations on the hyperoxia theme) is widely used to study PDR and other ischaemia-induced retinopathies, but it is not ideal. This is a neonatal model in which diabetes is absent and the knowledge acquired from the OIR needs to be carefully interpreted in the context of human DR.

OIR is not a diabetic model and an interesting variation has been recently reported in which the neonates are exposed to a degree of hyperglycaemia (Kermorvant-Duchemin et al. 2013). The Sennlaub group induced hyperglycaemia in newborn rat pups using STZ and there was resultant impairment of retinal vascular development, which was associated with recruitment of inflammatory macrophages (Kermorvant-Duchemin et al. 2013). Although not a diabetic model per se, this novel approach may provide an option to dissect aspects of angiogenesis in DR.

5.8 Techniques to Detect Retinal Lesion in Animal Models of Diabetes

As outlined above, large animals which have been diabetic for more than 5 years often develop retinal lesions that are similar to that of human NPDR. The larger eye ought to provide scope for evaluating changes using traditional clinical approaches such as fundus imaging, fluorescein angiography (FA) and OCT, although most of the published studies have been reliant on post-mortem evaluations. In recent years, research groups have invested considerable effort in establishing 'clinical' imaging techniques to evaluate the living retinas in the small rodent eye. While detecting subclinical changes to the retina of small animals is challenging, over the past decade, there has been a rapid development in the field of small animal retinal imaging.

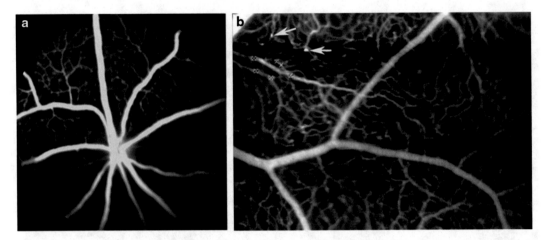

Fig. 5.7 **a** cSLO showing C57BL/6 mouse retinal blood vessels. **b** A representative image demonstrates that leukocytes trafficking in the retinal vessels (Courtesy of Professor Heping Xu, Queen's University Belfast). *Arrows*: leukocytes sticking inside the capillaries. Moving cells are shown in paired colour (moving from *red* to *blue* position)

5.8.1 Fundus Colour Imaging

With improved digital imaging technology, fundus images in rodents of excellent quality can now be readily obtained. However, since DR-related changes in rodent eyes are subclinical, the value of fundus images in rodent DR models is limited.

5.8.2 Confocal Scanning Laser Ophthalmoscope (cSLO)

Many of the DR-related changes in rodent models are related to retinal vascular damage. The cSLO is a powerful tool to investigate retinal vascular abnormalities. The cSLO system allows FFA to be performed by injecting sodium fluorescein into the animal either intravenously or intraperitoneally. Although the imaging technique is highly sensitive and reliable to detect BRB leakage in DR patients, it is less sensitive in diabetes rodents and not normally used to study BRB breakdown. Nevertheless, this technique is useful to study leukocyte–endothelial cell interaction (e.g. leukostasis) in diabetes (Serra et al. 2012). Leukocytes are labelled in vitro with a fluorescent dye (e.g. calcein AM), and are then adaptively transferred into a diabetic animal with a low concentration of sodium fluorescein. Leukocyte–endothelial interaction can be imaged in live animals (Fig. 5.7).

5.8.3 Optical Coherence Tomography (OCT)

Over the past decade, the OCT technology has developed rapidly in the ophthalmology field and micrometre resolution images can now be collected from human eyes. The technology acquires cross-sectional images from a series of laterally adjacent depth-scans, which can then be reconstructed into two-dimensional (2-D) or three-dimensional (3-D) images for analysis. OCT is extremely useful in diagnosing DMO in patients. A number of companies have modified the system for rodent eyes imaging. OCT can be used to measure retinal thickness as an amalgam of the different layers of the retina as well as the overall thickness. Post-mortem follow-up of OCT measurements of retinal layer thickness and comparison with retinal histology in rodents shows the relevance of the in vivo imaging approach. Unlike DR patients who often experience increased retinal thickness due to oedema, diabetes mouse or rat may present *reduced* retinal thickness after 6–9 months of diabetes duration which is related to progressive loss of neurons and/or glia in the neuropile (Hombrebueno et al. 2014). Figure 5.8

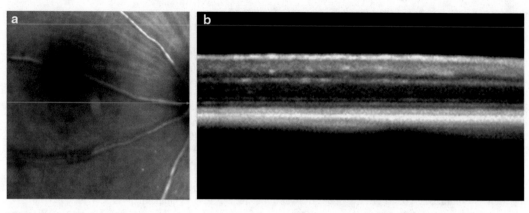

Fig. 5.8 SD-OCT image from a 3-month-old C57BL/6 diabetic mice acquired using Heidelberg SD-OCT. **a** HRA2 image and **b** OCT image

shows the OCT images of a 3-monthdiabetic mouse. No visible morphological changes can be detected at this time point (Authors' unpublished observation).

5.8.4 Electroretinogarphy (ERG)

ERG measures the electrophysiological response of the retina to light stimuli. ERG is not normally a routine examination for DR patients, perhaps due to technical and time demands and the fact that most ophthalmologists remain solely focused on the vascular component of the disease. In experimental studies of DR in animal models, ERG has been widely conducted, and it has been shown that diabetes induces a reduced a- and b-wave amplitude and prolonged implicit time in the oscillatory potentials (Ops; Hombrebueno et al. 2014). These changes are detected in both diabetic rats (Kohzaki et al. 2008) and mice (Hombrebueno et al. 2014).

5.9 Conclusion

Although various diabetes models have been developed, only large animals or primates with prolonged diabetes duration develop retinal lesions similar to clinical classifications of human DR. While diabetic mice and rats only develop early retinal neural and vascular changes, there is a clear merit in evaluating the basic pathogenesis of DR, uncovering new molecular targets and evaluating drug safety/efficacy prior to entry into phase I/II clinical trials. Therefore rodent diabetes models are set to remain the most popular animal models for research into DR. While all currently available models have limitations, they continue to have a critical value as we progress towards the development of efficacious new treatments for retinopathy in diabetic patients.

Compliance with Ethical Requirements Informed consent and animal studies disclosures are not applicable to this review.

Dr Chen, Professor Stitt and Prof. Lois declare that they have no conflict of interest.

References

Abari E, Kociok N, Hartmann U, Semkova I, Paulsson M, Lo A, Joussen AM (2013) Alterations in basement membrane immunoreactivity of the diabetic retina in three diabetic mouse models. Graefe's archive for clinical and experimental ophthalmology = Albrecht von Graefes Archiv fur klinische und experimentelle Ophthalmologie 251 (3):763–75

Ali Rahman IS, Li CR, Lai CM, Rakoczy EP (2011) In vivo monitoring of VEGF-induced retinal damage in the Kimba mouse model of retinal neovascularization. Curr Eye Res 36(7):654–62

Alvarez Y, Chen K, Reynolds AL, Waghorne N, O'Connor JJ, Kennedy BN (2010) Predominant cone photoreceptor dysfunction in a hyperglycaemic model of non-proliferative diabetic retinopathy. Dis Model Mech 3(3–4):236–45

Anderson HR, Stitt AW, Gardiner TA, Lloyd SJ, Archer DB (1993) Induction of alloxan/streptozotocin diabetes in dogs: a revised experimental technique. Lab Anim 27(3):281–5

Antonetti DA, Klein R, Gardner TW (2012) Diabetic retinopathy. N Engl J Med 366(13):1227–39

Barber AJ, Antonetti DA, Kern TS, Reiter CE, Soans RS, Krady JK, Levison SW, Gardner TW, Bronson SK (2005) The Ins2Akita mouse as a model of early retinal complications in diabetes. Invest Ophthalmol Vis Sci 46(6):2210–8

Berner AK, Brouwers O, Pringle R, Klaassen I, Colhoun L, McVicar C, Brockbank S, Curry JW, Miyata T, Brownlee M, Schlingemann RO, Schalkwijk C, Stitt AW (2012) Protection against methylglyoxal-derived AGEs by regulation of glyoxalase1 prevents retinal neuroglial and vasodegenerative pathology. Diabetologia 55(3):845–54

Bogdanov P, Corraliza L, Villena JA, Carvalho AR, Garcia-Arumi J, Ramos D, Ruberte J, Simo R, Hernandez C (2014) The db/db mouse: a useful model for the study of diabetic retinal neurodegeneration. PLoS ONE 9(5):e97302

Brownlee M (2001) Biochemistry and molecular cell biology of diabetic complications. Nature 414(6865):813–20

Buchi ER, Kurosawa A, Tso MO (1996) Retinopathy in diabetic hypertensive monkeys: a pathologic study. Graefe's archive for clinical and experimental ophthalmology = Albrecht von Graefes Archiv fur klinische und experimentelle Ophthalmologie 234(6):388–98

Connor KM, Krah NM, Dennison RJ, Aderman CM, Chen J, Guerin KI, Sapieha P, Stahl A, Willett KL, Smith LE (2009) Quantification of oxygen-induced retinopathy in the mouse: a model of vessel loss, vessel regrowth and pathological angiogenesis. Nat Protoc 4(11):1565–73

Cox O, Stitt AW, Simpson DA, Gardiner TA (2003) Sources of PDGF expression in murine retina and the effect of short-term diabetes. Mol Vis 10(9):665–72

Cusick M, Chew EY, Ferris F 3rd, Cox TA, Chan CC, Kador PF (2003) Effects of aldose reductase inhibitors and galactose withdrawal on fluorescein angiographic lesions in galactose-fed dogs. Arch Ophthalmol 121(12):1745–51

DCCT (1993) The effect of intensive treatment of diabetes on the development and progression of long-term complications in insulin-dependent diabetes mellitus. The Diabetes Control and Complications Trial Research Group. N Engl J Med 329(14):977–86

Engerman RL, Kern TS (1984) Experimental galactosemia produces diabetic-like retinopathy. Diabetes 33(1):97–100

Engerman RL, Kern TS (1987) Progression of incipient diabetic retinopathy during good glycemic control. Diabetes 36(7):808–12

Engerman RL, Kern TS, Larson ME (1990) Nerve conduction velocity in dogs is reduced by diabetes and not by galactosemia. Metabolism 39(6):638–40

ETDRS (1991) Early photocoagulation for diabetic retinopathy. ETDRS report number 9. Early Treatment Diabetic Retinopathy Study Research Group. Ophthalmology 98(5Suppl):766–785

Gardiner TA, Stitt AW, Anderson HR, Archer DB (1994) Selective loss of vascular smooth muscle cells in the retinal microcirculation of diabetic dogs. Br J Ophthalmol 78(1):54–60

Gardiner T, Gibson D, deGooyer T, Stitt A (2003a) Modulation of retinal angiogenesis in oxygen-induced retinopathy by inhibition of inflammatory cytokines. Am J Pathol, in press

Gardiner TA, Anderson HR, Degenhardt T, Thorpe SR, Baynes JW, Archer DB, Stitt AW (2003b) Prevention of retinal capillary basement membrane thickening in diabetic dogs by a non-steroidal anti-inflammatory drug. Diabetologia 46(9):1269–1275

Gastinger MJ, Singh RS, Barber AJ (2006) Loss of cholinergic and dopaminergic amacrine cells in streptozotocin-diabetic rat and Ins2Akita-diabetic mouse retinas. Invest Ophthalmol Vis Sci 47(7):3143–50

Gebarowska D, Stitt AW, Gardiner TA, Harriott P, Greer B, Nelson J (2002) Synthetic peptides interacting with the 67-kd laminin receptor can reduce retinal ischemia and inhibit hypoxia-induced retinal neovascularization. Am J Pathol 160(1):307–13

Gleeson M, Connaughton V, Arneson LS (2007) Induction of hyperglycaemia in zebrafish (Daniorerio) leads to morphological changes in the retina. Acta Diabetol 44(3):157–63

Hainsworth DP, Katz ML, Sanders DA, Sanders DN, Wright EJ, Sturek M (2002) Retinal capillary basement membrane thickening in a porcine model of diabetes mellitus. Comp Med 52(6):523–29

Hatchell DL, Toth CA, Barden CA, Saloupis P (1995) Diabetic retinopathy in a cat. Exp Eye Res 60(5):591–93

Hombrebueno JR, Chen M, Penalva RG, Xu H (2014) Loss of synaptic connectivity, particularly in second order neurons is a key feature of diabetic retinal neuropathy in the Ins2Akita mouse. PLoS ONE 9(5):e97970

Howell SJ, Mekhail MN, Azem R, Ward NL, Kern TS (2013) Degeneration of retinal ganglion cells in diabetic dogs and mice: relationship to glycemic control and retinal capillary degeneration. Mol Vis 19:1413–21

Intine RV, Olsen AS, Sarras MP Jr. (2013) A zebrafish model of diabetes mellitus and metabolic memory. J Vis Exp (72):e50232

Johnson MA, Lutty GA, McLeod DS, Otsuji T, Flower RW, Sandagar G, Alexander T, Steidl SM, Hansen BC (2005) Ocular structure and function in an aged monkey with spontaneous diabetes mellitus. Exp Eye Res 80(1):37–42

Joussen AM, Smyth N, Niessen C (2007) Pathophysiology of diabetic macular edema. Dev Ophthalmol 39:1–12

Kador PF, Akagi Y, Takahashi Y, Ikebe H, Wyman M, Kinoshita JH (1990) Prevention of retinal vessel changes associated with diabetic retinopathy in galactose-fed dogs by aldose reductase inhibitors. Arch Ophthalmol 108(9):1301–09

Kador PF, Takahashi Y, Sato S, Wyman M (1994) Amelioration of diabetes-like retinal changes in galactose-fed dogs. Prev Med 23(5):717–21

Kador PF, Takahashi Y, Akagi Y, Neuenschwander H, Greentree W, Lackner P, Blessing K, Wyman M

(2002) Effect of galactose diet removal on the progression of retinal vessel changes in galactose-fed dogs. Invest Ophthalmol Vis Sci 43(6):1916–21

Kermorvant-Duchemin E, Pinel AC, Lavalette S, Lenne D, Raoul W, Calippe B, Behar-Cohen F, Sahel JA, Guillonneau X, Sennlaub F (2013) Neonatal hyperglycemia inhibits angiogenesis and induces inflammation and neuronal degeneration in the retina. PLoS ONE 8(11):e79545

Kern TS, Engerman RL (1994) Comparison of retinal lesions in alloxan-diabetic rats and galactose-fed rats. Curr Eye Res 13(12):863–7

Kern TS, Engerman RL (1996) Capillary lesions develop in retina rather than cerebral cortex in diabetes and experimental galactosemia. Arch Ophthalmol 114(3):306–10

Kern TS, Miller CM, Tang J, Du Y, Ball SL, Berti-Matera L (2010) Comparison of three strains of diabetic rats with respect to the rate at which retinopathy and tactile allodynia develop. Mol Vis 16:1629–39

Kim SY, Johnson MA, McLeod DS, Alexander T, Otsuji T, Steidl SM, Hansen BC, Lutty GA (2004) Retinopathy in monkeys with spontaneous type 2 diabetes. Invest Ophthalmol Vis Sci 45(12):4543–53

King JL, Mason JO 3rd, Cartner SC, Guidry C (2011) The influence of alloxan-induced diabetes on Muller cell contraction-promoting activities in vitreous. Invest Ophthalmol Vis Sci 52(10):7485–91

Klein R, Klein BE, Moss SE, Davis MD, DeMets DL (1984) The Wisconsin epidemiologic study of diabetic retinopathy. II. Prevalence and risk of diabetic retinopathy when age at diagnosis is less than 30 years. Arch Ophthalmol 102(4):520–6

Kohzaki K, Vingrys AJ, Bui BV (2008) Early inner retinal dysfunction in streptozotocin-induced diabetic rats. Invest Ophthalmol Vis Sci 49(8):3595–3604

Kowluru RA, Engerman RL, Kern TS (2000) Abnormalities of retinal metabolism in diabetes or experimental galactosemia VIII. Prevention by aminoguanidine. Curr Eye Res 21(4):814–819

Lai AK, Lo AC (2013) Animal models of diabetic retinopathy: summary and comparison. J Diabetes Res 2013:106594

Lee SE, Ma W, Rattigan EM, Aleshin A, Chen L, Johnson LL, D'Agati VD, Schmidt AM, Barile GR (2010) Ultrastructural features of retinal capillary basement membrane thickening in diabetic swine. Ultrastruct Pathol 34(1):35–41

Mansour SZ, Hatchell DL, Chandler D, Saloupis P, Hatchell MC (1990) Reduction of basement membrane thickening in diabetic cat retina by sulindac. Invest Ophthalmol Vis Sci 31(3):457–63

McLenachan S, Chen X, McMenamin PG, Rakoczy EP (2013) Absence of clinical correlates of diabetic retinopathy in the Ins2Akita retina. Clin Exp Ophthalmol 41(6):582–92

McVicar CM, Hamilton R, Colhoun LM, Gardiner TA, Brines M, Cerami A, Stitt AW (2011) Intervention with an erythropoietin-derived peptide protects against neuroglial and vascular degeneration during diabetic retinopathy. Diabetes 60(11):2995–3005

Medina RJ, O'Neill CL, Humphreys MW, Gardiner TA, Stitt AW (2010) Outgrowth endothelial cells: characterization and their potential for reversing ischemic retinopathy. Invest Ophthalmol Vis Sci 51(11):5906–13

Medina RJ, O'Neill CL, O'Doherty TM, Knott H, Guduric-Fuchs J, Gardiner TA, Stitt AW (2011) Myeloid angiogenic cells act as alternative M2 macrophages and modulate angiogenesis through interleukin-8. Mol Med 17(9–10):1045–55

Olsen AS, Sarras MP Jr, Intine RV (2010) Limb regeneration is impaired in an adult zebrafish model of diabetes mellitus. Wound Repair Regen 18(5):532–42

Olsen AS, Sarras MP Jr, Leontovich A, Intine RV (2012) Heritable transmission of diabetic metabolic memory in zebrafish correlates with DNA hypomethylation and aberrant gene expression. Diabetes 61(2):485–91

Rakoczy EP, Ali Rahman IS, Binz N, Li CR, Vagaja NN, de Pinho M, Lai CM (2010) Characterization of a mouse model of hyperglycemia and retinal neovascularization. Am J Pathol 177(5):2659–70

Robinson R, Barathi VA, Chaurasia SS, Wong TY, Kern TS (2012) Update on animal models of diabetic retinopathy: from molecular approaches to mice and higher mammals. Dis Model Mech 5(4):444–56

Robison WG Jr, Kador PF, Kinoshita JH (1983) Retinal capillaries: basement membrane thickening by galactosemia prevented with aldose reductase inhibitor. Science 221(4616):1177–9

Roy MS, Gunkel RD, Podgor MJ (1986) Color vision defects in early diabetic retinopathy. Arch Ophthalmol 104(2):225–8

Saidi T, Mbarek S, Omri S, Behar-Cohen F, Chaouacha-Chekir RB, Hicks D (2011) The sand rat, Psammomys obesus, develops type 2 diabetic retinopathy similar to humans. Invest Ophthalmol Vis Sci 52(12):8993–9004

Sarras MP Jr, Leontovich AA, Olsen AS, Intine RV (2013) Impaired tissue regeneration corresponds with altered expression of developmental genes that persists in the metabolic memory state of diabetic zebrafish. Wound Repair Regen 21(2):320–8

Serra AM, Waddell J, Mannivannan A, Xu H, Cotter M, Forrester JV (2012) CD11b+ bone marrow-derived monocytes are the major leukocyte subset responsible for retinal capillary leukostasis in experimental diabetes in the mouse and express high levels of CCR5 in the circulation. Am J Pathol 181(2):719–27

Simpson DA, Murphy GM, Bhaduri T, Gardiner TA, Archer DB, Stitt AW (1999) Expression of the VEGF gene family during retinal vaso-obliteration and hypoxia. Biochem Biophys Res Commun 262(2):333–40

Smith LE, Wesolowski E, McLellan A, Kostyk SK, D'Amato R, Sullivan R, D'Amore PA (1994) Oxygen-induced retinopathy in the mouse. Invest Ophthalmol Vis Sci 35(1):101–11

Sokol S, Moskowitz A, Skarf B, Evans R, Molitch M, Senior B (1985) Contrast sensitivity in diabetics with and without background retinopathy. Arch Ophthalmol 103(1):51–4

Stitt AW, Lois N, Medina RJ, Adamson P, Curtis TM (2013) Advances in our understanding of diabetic retinopathy. Clin Sci (Lond) 125(1):1–17

Tang J, Kern TS (2011) Inflammation in diabetic retinopathy. Prog Retin Eye Res 30(5):343–58

The Diabetes C, Complications Trial/Epidemiology of Diabetes I, Complications Research G (2014) Effect of intensive diabetes therapy on the progression of diabetic retinopathy in patients with type 1 diabetes: 18years of follow-up in the DCCT/EDIC. Diabetes. doi:10.2337/db14-0930

Tso MO, Kurosawa A, Benhamou E, Bauman A, Jeffrey J, Jonasson O (1988) Microangiopathic retinopathy in experimental diabetic monkeys. Trans Am Ophthalmol Soc 86:389–421

Tzekov R, Arden GB (1999) Theelectroretinogram in diabetic retinopathy. Surv Ophthalmol 44(1):53–60

(UKPDS) UPDS (1998) Effect of intensive blood-glucose control with metformin on complications in overweight patients with type 2 diabetes (UKPDS 34). The Lancet 352(9131):854–65

van Eeden PE, Tee LB, Lukehurst S, Lai CM, Rakoczy EP, Beazley LD, Dunlop SA (2006) Early vascular and neuronal changes in a VEGF transgenic mouse model of retinal neovascularization. Invest Ophthalmol Vis Sci 47(10):4638–45

Vessey KA, Wilkinson-Berka JL, Fletcher EL (2011) Characterization of retinal function and glial cell response in a mouse model of oxygen-induced retinopathy. J Comparative Neurol 519(3):506–27

Winzell MS, Ahren B (2004) The high-fat diet-fed mouse: a model for studying mechanisms and treatment of impaired glucose tolerance and type 2 diabetes. Diabetes 53(Suppl3):S215–9

Wisniewska-Kruk J, Klaassen I, Vogels IM, Magno AL, Lai CM, Van Noorden CJ, Schlingemann RO, Rakoczy EP (2014) Molecular analysis of blood-retinal barrier loss in the Akimba mouse, a model of advanced diabetic retinopathy. Exp Eye Res 122:123–31

Zhang SX, Ma JX, Sima J, Chen Y, Hu MS, Ottlecz A, Lambrou GN (2005) Genetic difference in susceptibility to the blood-retina barrier breakdown in diabetes and oxygen-induced retinopathy. Am J Pathol 166(1):313–21

Commentary

Noemi Lois

e-mail: n.lois@qub.ac.uk

Centre for Experimental Medicine,

Queen's University,

Belfast, UK

Despite of early detection, by means of diabetic retinopathy (DR) screening programmes, improved metabolic control and new therapies, DR remains a leading cause of visual impairment and blindness worldwide. In DR sight loss occurs as a result of the development of diabetic macular edema (DME), ischaemic maculopathy and/ or proliferative diabetic retinopathy (PDR). Imperfect treatments are available for these complications once established; there is no therapeutic intervention to prevent their development. Thus, the search to improve the care of people with DR must continue.

Animal models of disease are extremely valuable, if mimicking adequately the disease in humans, to investigate pathogenic pathways involved and, subsequently, design new therapeutic strategies. It is essential for researchers and clinicians to understand well the nature of the disease and the mechanisms involved in its development in these animal models to interpret correctly any results generated and their relevance to human disease. In this chapter, Drs Chen and Stitt (1) provide us with a thorough review of the experimental animal models of DR available, pointing out their advantages and shortcomings.

It is apparent that animal models reproduce early features of DR. But there are additional requirements to take into consideration. For instance, the two histopathological lesions essentially unique to DR which, in addition, are the earliest observed, namely the diffuse thickening of the capillary basement membranes (somewhat different to that occurring in ageing) and the selective loss of pericytes, do occur in animal models and are often seen after 6-9 months duration of diabetes. The pathogenic routes involved in the occurrence of these early features can be, hence, investigated; therapeutic measures can be also sought. However, these features cannot be recognised by the in vivo imaging technologies available. To determine change in these parameters result of a potential intervention, terminal studies and, subsequently, the use of high number of animals would be required. Furthermore and even more importantly, translating findings into humans in potential phase 1 interventional trials would be challenging, as the endpoints evaluated in the pre-clinical studies (i.e. the thickening of the basement membrane and pericyte loss) would not be detectable in humans by current means. Thus, although the animal model may mimic closely the change observed in humans, the end points evaluated are not translatable to clinical studies.

Abnormalities occurring as a result of the above and other histopathological changes in early DR, for example increased vascular permeability and areas of non-perfusion, may be suitably imaged. However, at present time, quantitative evaluation on these endpoints remains challenging. Moreover, current imaging techniques to evaluate them are invasive, requiring an administration of a dye intravenously. Although fluorescein angiography is a test routinely used in Ophthalmology clinics, its invasive nature would still limit its use in human investigations.

Functional studies to determine early functional abnormalities in animal models of DR have been scarce. These may be more difficult to undertake but have the potential value of being objective and more likely to be translatable to human studies. Furthermore, they may be more significant to people with DR as it is the functional loss, rather than the anatomical change, what matters to those affected by the disease and especially considering that structural changes do not relate always to reduce or loss of function.

Considerations may be given by basic scientists to the above matters when designing pre-clinical studies using animal models of DR. Endpoints that can be obtained by using reproducible, user-friendly, non-invasive technologies should be favoured.

Lastly, and also as discussed by Chen and Stitt in their comprehensive review, there is no adequate animal model of PDR. PDR may develop and progress in humans very quickly. Large oscillations in glucose levels, a rapid and tight glucose control after a period of chronically high levels of glycemia, kidney failure, etc, may precipitate its fast development and progression. It is possible that, in order to achieve development of PDR in experimental animal models these events may need to be mimicked; combination of insults (for example, high glucose levels and blood pressure within the context of a genetically predisposed strain) may be also required.

It is clear that work must continue. Basic scientists and clinicians should work together in the search for improved in vivo models and endpoints for research into DR with the final goal of improving the quality of life of people with DR.

Animal Models of Autoimmune Uveitis

Jennifer L. Kielczewski and Rachel R. Caspi

6.1 Introduction

Uveitis is a heterogeneous group of intraocular inflammatory eye diseases, which can lead to irreversible vision loss and blindness (Srivastava et al. 2010). The eye is a unique organ with a distinct blood–retinal barrier (BRB), which provides it with an immune privileged status (Streilein 2003a, b; Simpson 2006). This means that the eye should not be subjected to immune attack or invasion. However, breakdown of immune privilege and the BRB can occur, leading to ocular inflammation known as uveitis (Caspi 2006). This ocular disease is classified as anterior, intermediate, posterior, or panuveitis, depending on the location of ocular tissues involved (Lee et al. 2014). The etiology of uveitis can be infectious and noninfectious, but is idiopathic in approximately 30–50 % of cases (Lyon et al. 2009; Forrester et al. 2013; Lee et al. 2014). Infectious causes of uveitis include those limited to the eye resulting from bacterial (tuberculosis), parasitic (toxoplasmosis), viral (HSV/VSV and CMV), or fungal

agents (Candida; Rothova et al. 1992; Rothova et al. 1996; Suttorp-Schulten and Rothova 1996). Often, uveitis has an autoimmune component that can be linked to underlying systemic diseases such as Reiter's syndrome, sarcoidosis, Behcet's syndrome, multiple sclerosis, inflammatory bowel disease, or Vogt–Koyanagi–Harada syndrome (Levy et al. 2011). Uveitis related to autoimmune disease is more common in developed countries, while infectious uveitis is prevalent in underdeveloped countries, where infectious diseases can be more rampant (Lee et al. 2014). Genetic associations, such as HLA B27 in ankylosing spondylitis, or HLA A29 in birdshot chorioretinopathy (BC), have also been associated with pathogenesis of uveitis (Lyon et al. 2009). Uveitis cases in the USA total about 150,000 per year and are estimated to cause approximately 10 % of vision loss or blindness (Caspi 2010; Gritz and Wong 2004). It is a potentially blinding ocular disease that can have significant socioeconomic consequences when left untreated or poorly managed. New therapies are needed to better control and treat ocular inflammation associated with uveitis, particularly the ones that are administered by localized delivery directly into the eye.

6.2 Why Do We Need Animal Models to Study Human Uveitis?

Conventional therapies for uveitis rely on general immunosuppressive therapies such as corticosteroids, often in combination with cyclosporin,

R. R. Caspi (✉)
Laboratory of Immunology, National Eye Institute, National Institutes of Health, 10 Center Drive, Building 10, Room 10N222, Bethesda, MD 20892–1857, USA
e-mail: rcaspi@helix.nih.gov

J. L. Kielczewski
Laboratory of Immunology, National Eye Institute at National Institutes of Health, Bethesda, MD 20892, USA
e-mail: nussenblattr@nei.nih.gov

rapamycin, or similar agents. Immunosuppressive drugs can have serious side effects and some patients are not responsive to these conventional treatments. Understanding the basic inflammatory mechanisms of uveitis is a prerequisite to the development of new therapies and treatments. To better understand the pathogenesis of uveitis and to develop more effective treatment modalities, animal models are needed to mimic the human disease state (Horai and Caspi 2011; Caspi 2011). Ideally, well-characterized human uveitic or inflamed tissue would be desired for research. However, due to ethical issues connected to experimentation in human subjects and limited eye bank donor tissue, animal models are deemed necessary to study uveitis (Lyon et al. 2009).

Animal models are specifically needed to dissect out T cell effector Th1 and Th17 cellular responses, both of which are associated with clinical uveitis. Animal models have shown us naïve T cells that mature into either Th1 (IFN-γ producing) or Th17 (IL-17 producing) effector cells find their way into the ocular environment upon breach of the BRB. Upon recognition of specific antigens within the eye, they orchestrate a destructive inflammatory process by recruiting inflammatory leukocytes from the circulation. Understanding how T cell effector responses initiate cellular cascade signaling mechanisms (both innate and adaptive responses) and cytokine production in the eye will have direct implications in the design of novel therapies for uveitis.

6.3 Animal Models of Uveitis Exhibit Clinical Heterogeneity, Similar to Human Uveitis

Several experimental animal models of noninfectious uveitis have been developed. The animal models develop ocular inflammation similar to that found in human uveitis (Agarwal and Caspi 2004). Animal models often display similar features to human clinical cases such as retinal and choroidal inflammation, vasculitis, retinal atrophy, and ultimately loss of vision (Levy et al. 2011). Often, animals display heterogeneity in terms of their inflammatory response much like

humans (Caspi et al. 2008). For example, animals of the same species do not necessarily respond the same way to the same antigen. In rats, both photoreceptor proteins, interphotoreceptor retinoid-binding protein (IRBP) and arrestin (S antigen), are highly immunogenic, but in most strains of mice the antigen that elicits pathology is IRBP (Agarwal and Caspi 2004). Likewise, different strains have varying degrees of susceptibility (Caspi et al. 2008). Mice of the B10.RIII strain is highly susceptible to experimentally induced uveitis, while the C57/BL6 mouse strain is less susceptible (Agarwal and Caspi 2004). Table 6.1 describes the three major categories of animal models of autoimmune uveitis, which include induced, spontaneous, and humanized animal models. The specific animal models encompassing each category will be discussed in detail, highlighting their significance for the study of ocular inflammation and uveitis, as well as discussing their limitations.

6.4 Induced Animal Models of Uveitis

6.4.1 Experimental Autoimmune Uveitis (EAU)

Over the span of five decades, different animal models of induced uveitis have been developed. The most common model is the "classical" EAU with complete Freund's adjuvant (CFA). This model was originally developed by Aronson et al. in 1963, in the guinea pig using homologous uveal tissue, and was subsequently adapted to the rat using photoreceptor extracts by Wacker and Kaslow in 1973. In 1981, Kozak et al. refined the rat model using a purified retinal soluble antigen (S-Ag). EAU has subsequently been induced in the rat with IRBP, RPE-65, rhodopsin, recoverin, and phosducin retinal-specific proteins (Gery et al. 1986; Caspi et al. 1988). The rat model has shown that different retinal proteins induce disease and that the disease is T cell mediated and transferable with long-term CD4$^+$ T cell lines specific to retinal antigens. In 1988, successful induction of EAU in mice was achieved by

Table 6.1 Major animal models of uveitis

Type of model	Model description	T cell mediated response	Advantages and disadvantages
1. EAU induced animal models of uveitis			
A. "Classical" EAU: immunization of wild-type mice with an ocular antigen in CFA. Also, can adoptively transfer immune cells or pathogenic cell line from immunized donors to naive recipients	EAU induced in B10.RIII, B10.A and C57BL/6 mice with IRBP	Classical EAU is Th17 driven Adoptive transfer of polarized retina specific effector T cells is Th1 or Th17 driven	Mycobacteria in CFA provide innate "danger" signals that can polarize autoimmune lymphocytes towards a pro-inflammatory phenotype. Pertussis toxin (PTX) is needed as an additional adjuvant in less susceptible strains. Model is highly reproducible and provides consistent disease similar to human disease. Mice can be used in preclinical studies and testing of potential therapeutic agents
B. Immunization of rats with retinal antigen in CFA or adoptive transfer of T cells from immunized donors to naive recipients	Arrestin-induced model in the Lewis rat. Other retinal and choroidal (melanin) antigens are uveitogenic in Lewis rats	Th17 driven	Was the major EAU model until 1988. Currently, it is not often used
C. EAU by infusion of antigen-pulsed syngenic dendritic cells (DCs)	B10.RIII mice are given splenic DCs elicited with Flt3, matured in vitro, and pulsed with IRBP p 161–180	DC-EAU is Th1 driven	Requires two injections of DCs and pertussis toxin. Less severe disease than CFA-EAU and depends mainly on Th1 cells
D. Neo-self-antigen expressed in the mouse eye transgenically under the control of an eye-specific promoter or can be retrovirally introduced influenza hemagglutin (HA) in the retina	Transgenic HEL or β-gal under the control of an eye-specific promoter or by retroviral transduction following intraocular injection. Mice are then immunized with the specific antigen	Th1 driven	Uveitogenic T cells need to be activated
2. Humanized animal models of uveitis			
A. "Humanized" EAU: HLA-DR3 transgenic mice are immunized with an ocular antigen or adoptively transferred with T cells from immunized donors	Induced EAU in HLA-DR3 transgenic mice immunized with retinal arrestin or its peptide fragments	Not well characterized	Mycobacteria in CFA provide innate "danger" signals that can polarize autoimmune lymphocytes towards a pro-inflammatory phenotype
B. Mice transgenic for a human HLA class I antigen associated with uveitis	HLA-A29 transgenic mice. Birdshot retinochoroidopathy like disease develop between 8–12 months of age	Not well characterized	Mice require "aging" for disease to manifest. Model is clinically relevant

Table 6.1 (continued)

Type of model	Model description	T cell mediated response	Advantages and disadvantages
3. Spontaneous animal models of uveitis			
A. AIRE knockout mice	Spontenous uveitis occurs in AIRE-deficient mice directed at IRBP	Not well characterized	High frequency and affinity of cells bear TCRs specific for IRBP. May not represent the physiological state found in humans. Not much is known about the effector T cell type.
B. EAU in IRBP TCR trangenic mice	Spontaneous uveitis occurs in mice by weaning age. There are three lines with high (R161H), medium (R161M), and low (R161L) disease	EAU in R161 mice is Th1 driven	Mice can be used to study natural triggers of uveitis. There is no need for CFA or an adjuvant. Preclinical studies can be done in the three different lines, which produce different severities of disease. Mice can develop lymphomas
C. Double-transgenic mice expressing a neoself-antigen in the retina and an antigen-specific TCR	Spontaneous EAU-like uveitis in mice expressing ocular HEL and a HEL-specific TCR	Th1 driven	Spontaneous nature of disease in this model is dependent on the frequency and affinity of the TCR

IRBP interphotoreceptor retinoid-binding protein, *CFA* complete Freund's adjuvant, *AIRE* autoimmune regulator

Caspi et al. using the whole IRBP protein (Caspi et al. 1988). S-Ag was less potent than IRBP as a pathogenic antigen in mice, and in fact, mouse strains that developed EAU after immunization with IRBP failed to do so after immunization with S-Ag. Since then, pathogenic epitopes of the IRBP protein have been characterized, such as peptides 1–20, 161–180, and 201–216 that elicit EAU in mice (Caspi 2003). EAU can also be induced in rabbits and monkeys, which have larger eyes, making them well suited for experimental manipulation (Zeiss 2013). The EAU model, however, is most frequently used in mice and rats (Agarwal and Caspi 2004). Mice are particularly useful due to their immunologically and genetically well-defined nature and ease of genetic manipulability (Zeiss 2013). There is a vast array of murine transgenic strains and knockouts for immunologically relevant genes. Therefore, the rat EAU model induced with retinal arrestin (retinal soluble antigen, S-Ag) or with IRBP and the mouse EAU model induced with IRBP are the most commonly used EAU models (Agarwal and Caspi 2004; Wildner et al. 2008; Horai and Caspi 2011).

6.4.2 Different Uveitogenic Antigens Induce Different Variants of Uveitis

Researchers induce EAU in rodents using a single dose (50–100 μg) of soluble retinal-specific antigen like bovine IRBP (whole IRBP protein or a pathogenic IRBP peptide fragment like peptides 161–180 or 1–20 can be used) emulsified in CFA, supplemented with heat-killed mycobacteria (Caspi 2008a; Agarwal and Caspi 2004) needed to activate the innate system in order to boost the immune response. Also, injection of *Bordetella pertussis toxin* (PTX) as an additional adjuvant at the time of subcutaneous immunization is required in less susceptible mouse strains (Silver et al. 1999). The adjuvant effect of PTX is complex and not completely understood. It is thought to contribute to inhibition of lymphocyte recirculation, stimulatory effects on B and T cells, and enhancement of vascular permeability

and disruption of the blood–organ barrier (Caspi 2003). Nevertheless, rodents develop intraocular inflammation 9–12 days after immunization, and the inflammation typically subsides by 21–28 days (Agarwal and Caspi 2004). The clinical course of the disease is typically monophasic and acute. Antigen dose and the site of immunization impact EAU disease severity, and the antigen can be easily increased or decreased depending on the desired disease range. Disease scores are assessed by microscopic fundus examination or histopathology using a scale of 0 (no disease) to 4 (severe disease). Table 6.2 consists of additional details on EAU scoring. Moderate to severely diseased animals can develop subretinal/chorioretinal granulomas along with retinal vasculitis,

Table 6.2 Scoring of EAU in the mouse by histopathology and fundoscopy

Disease score	Histopathoiogical characteristics	Fundoscopy characteristics
0	No disease, normal retinal structure	No change
0.5 (trace)	Mild inflammatory cell infiltration, no tissue damage	Few peripheral focal lesions, minimal vasculitis
1	Infiltrations into vitreous, retina, retinal folding and vasculitis, one granuloma	Mild vasculitis, less than 5 small focal lesions
2	Moderate infiltration into the vitreous, uvea and retina, retinal folds and detachment, focal photoreceptor damage, small granulomas	Multiple chorioretinal lesions and/or infiltrations, severe vasculitis
3	Moderate-to-severe infiltration into uvea, vitreous and retina, extensive retinal folding with retinal detachment, subretinal neovascularization, medium granulomas	Large lesions, subretinal neovascularization, retinal hemorrhages
4	Severe infiltration, subretinal neovascularization and hemorrhages, extensive photoreceptor damage, large granulomatous lesions	Large retinal detachment, retinal atrophy

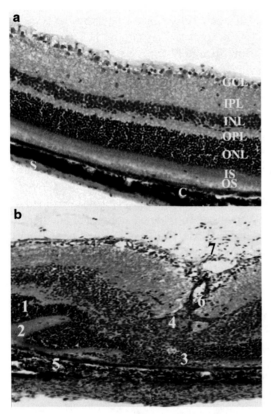

Fig. 6.1 Histopathology showing the cardinal signs of uveitis in mouse retina, which is similar to that reported in human uveitis. **a** Represents a normal mouse retina with all the intact retinal layers. **b** Shows an R161H mouse with spontaneous uveitis. There is significant inflammation and disruption of the retinal integrity. *1* Retinal folds, *2* retinal detachment, *3* photoreceptor loss, *4* granuloma formation, *5* choroiditis, *6* vasculitis (retinal vessel inflammation), *7* vitritis (a sign of active inflammatory activity). All of these pathological features can occur in varying degrees of severity and extent. *GC* ganglion cell layer, *IPL* inner plexiform layer, *OPL* outer plexiform layer, *ONL* outer nuclear layer, *PR* photoreceptor layer, *C* choroid, *S* sclera

retinal and optic nerve edema, photoreceptor loss, and vitreous inflammatory cell infiltration (macrophages and lymphocytes), which are all clinical features seen in human uveitis. Figure 6.1 shows some of the pathological characteristics seen in mice with EAU. As mentioned earlier, susceptibility to disease is strain-dependent. In mice and rats, both major histocompatibility complex (MHC) and non-MHC genes have effects on disease susceptibility (Pennesi and Caspi 2002;

Mattapallil et al. 2008). MHC control is likely due to the ability to bind and present uveitogenic epitopes. The Lewis rat is the most commonly used rat strain for EAU (Wildner et al. 2008). In the mouse, the most highly susceptible strain so far is B10.RIII, followed by B10.A strain, and then C57BL/6, which are only moderately susceptible (Agarwal and Caspi 2004; Caspi 2008b). In this model, both IFN-γ and IL-17A producing cells are detected in inflamed eyes, but CFA EAU is prevented and reversed with anti-IL-17A antibodies, indicating its dependence on the Th17 effector lineage (Horai and Caspi 2011).

6.4.3 Experimental Melanin Protein-Induced Uveitis (EMIU)

It is important to note that non-retinal proteins can also be used to induce uveitis in animals, examples include melanin or tyrosinase-related proteins, which are proteins in the melanin synthesis pathway, and even melanin basic protein (MBP; distinct from MBP = myelin basic protein that is used to elicit experimental autoimmune encephalomyelitis-associated anterior uveitis (EAE/AU) ahead; Chan et al. 1994; Bora et al. 1995; Commodaro et al. 2011). Each of these antigens can induce uveitis of somewhat different form and duration. For example, non-soluble melanin proteins isolated from pigmented bovine eye tissues can induce uveitis in a model known as EMIU, which was described in the 1990s by Broekhuyse et al. (Broekhuyse et al. 1991, 1992, 1991). This model can be induced by subcutaneous injection of melanin proteins in association with Hunter adjuvant in rats and monkeys. Ocular inflammation is evident in iris, ciliary body, and choroid. Ocular inflammation is observed 10–14 days after immunization and consists of conjunctival hyperemia, corneal edema, anterior uveitis, iris vessel dilation, and synechiae (Gasparin et al. 2012). Both CD4[+] and CD8[+] T cells, macrophages, and neutrophils infiltrate into the eye. IL-2 and IFN-γ are the major intraocular cytokines produced early in the disease, while IL-12 and TNF-α play a role later in the inflammatory response (Papotto et al. 2014; Gasparin et al. 2012). It has also been suggested that complement plays a role in the EMIU model (Jha et al. 2006). Essentially, this model is useful for studying anterior uveitis (iridocyclitis), where melanin is found, and its spontaneous recurrence (Smith et al. 2008). The relapsing feature of this model makes it useful for the evaluation of therapies initiated during established disease and aimed at the prevention of recurrence.

6.4.4 Experimental Autoimmune Encephalomyelitis-Associated Anterior Uveitis

Myelin basic protein can also induce inflammation in rat eyes following subcutaneous immunization with CFA (Gasparin et al. 2012). This experimental model is known as EAE/AU (Constantinescu and Lavi 2000) (Adamus et al. 1996). Mice and rats immunized with MBP will typically experience ocular inflammation 10–12 days after immunization. Clinical symptoms include iris vessel dilation, anterior uveitis, and synechiae, which can persist for 30 days (Adamus et al. 1996). IL-2 and IFN-γ produced by Th1 cells are the predominant inflammatory mediators present in this model (Gasparin et al. 2012; Adamus et al. 1996). This model is particularly ideal for studying anterior uveitis and the inflammatory relapses associated with this disease.

6.4.5 Endotoxin-Induced Uveitis (EIU)

Endotoxin can be used to induce ocular inflammation in a model known as EIU. This is a nonautoimmune model of uveitis. It can be induced by intravenous, intraperitoneal, or subcutaneous injection of low doses of endotoxin in mice and rats (Rosenbaum et al. 1980; Papotto et al. 2014). In rats, macroscopically ocular inflammation can appear within 18–24 h post-LPS injection and can resolve spontaneously within 2–3 days (Papotto et al. 2014). The histopathology of EIU is characterized by transient, but intense, acute inflammatory cellular infiltration of neutrophils and macrophages, as well as protein accumulation in the anterior chamber and mild posterior uveitis (Bora et al. 1995; Broekhuyse et al. 1992). In mice, inflammation is typically much milder and is best quantitated by cellular counts and/or measurement of cytokines in ocular fluids (Li et al. 1995). Various cytokines and chemokines are released by infiltrating and ocular resident cells, including TNF-α, IFN-γ, TGF-β, IL-1, IL-6, CCL-2, and CCL-5 (Gasparin et al. 2012). Repeated injections of LPS can result in a state of tolerance in animals. This model is useful to study acute and subacute anterior ocular inflammation in humans.

6.4.6 Adoptive Transfer Model of T Cells from EAU Challenged Animals

Immune T cells from EAU-induced animals can transfer disease to naïve, genetically compatible recipient animals by a process known as adoptive transfer (Caspi 2003). Primary cultures from immunized donors are used as a source of uveitogenic effector cells. T cells are isolated from lymph nodes and spleens of donors immunized with a retinal antigen, such as IRBP, and activated in vitro with the same immunizing ocular antigen (Agarwal et al. 2012; Caspi 2003). The activated effector T cells (typically 30–50 million) are then infused intravenously into host animals after 3 days in culture (Caspi 2003). The recipients of these cells develop a destructive disease, usually within 1 week. Loss of photoreceptor cells, infiltration of inflammatory cells, and loss of vision can be seen in the eye of recipient mice. Also, long-term antigen-specific T cell lines, which are typically CD4+ T cells of the Th1 phenotype, can be derived from draining lymph node cells of IRBP or peptide-immunized mice and transferred into the recipient mice (Agarwal and Caspi 2004; Caspi 2003, 2008c, 2010, 2011). The adoptive transfer model is useful to analyze effector mechanism(s) of disease. It avoids the use of adjuvants in the recipients; hence, it is considered by some to be more reminiscent of human clinical uveitis.

6.4.7 EAU Induced with Antigen-Pulsed Dendritic Cells (DCs)

DCs are professional antigen presenting cells (APCs) capable of stimulating naïve T cells and are likely the main APCs in the early stages of EAU induction. A new model of EAU was developed by injection of matured splenic DC loaded with the major uveitogenic peptide of IRBP into the B10.RIII strain of mice (Caspi 2008a; Tang et al. 2007). The duration of the disease is shorter, pathology is less severe, and the inflammatory infiltrate is different from the classical EAU model (Caspi et al. 2008). Fundus lesions are punctate, and the inflammatory infiltrate contains abundant granulocytes and relatively few mononuclear cells. Notably, EAU resulting from antigen-pulsed DCs is not only clinically distinct from classical EAU, but also driven by different effector mechanisms (Caspi 2008a). DC EAU cannot be induced by infusion of uveitogenic DC in IFN-γ deficient recipients, indicating a dependence on the Th1 effector lineage rather than Th17 like CFA EAU. This model may represent some types of uveitis that are not adequately represented by the classical EAU model and may offer insights into the heterogeneous nature of human disease (Horai and Caspi 2011).

6.4.8 Other EAU-Like Models

Finally, neoantigens expressed in the retina through genetic engineering can serve as target antigens for uveitis. These foreign proteins include hen-egg lysozyme (HEL) or β-galactosidase (β-gal) that have been transgenically expressed in the retina or lens under the control of a tissue-specific promoter to serve as a neo-self-antigen. Gregerson et al. induced expression of the foreign protein β-galactosidase (β-gal) in the mouse retina and demonstrated that EAU can be induced by immunizing mice with β-gal. Alternatively, retinal cells can also be transduced in vivo to express a foreign-protein-like influenza hemagglutinin (HA), using a viral vector (Terrada et al. 2006). Subsequent immunization with the neoantigen, or transfer of immune cells from genetically compatible donors, triggers disease in the eye, similarly to immunization with IRBP or adoptive transfer of IRBP-specific T cells in wild-type mice (Gregerson et al. 1999; Lai et al. 1999; Terrada et al. 2006; Lambe et al. 2007).

Some neoantigen-driven models can result in spontaneous uveitis, specifically, HEL in combination with the HEL-specific 3A9 TCR (Lai et al. 1999; Lambe et al. 2007), and these are discussed below in the spontaneous animal models section.

6.4.9 The Future of EAU- and EAU-Like Induced Models in Translational Research

Various EAU models are used to represent human uveitic diseases of a putative autoimmune nature, such as sympathetic ophthalmia, birdshot retinochoroidopathy, and Behcet's disease (Bodaghi and Rao 2008; Forrester et al. 2013; Nussenblatt 2002). Although the EAU models reproduce some of the same features in human uveitis, they are not perfect. We have not definitively identified which retinal antigens are involved in human uveitis. However, the fact that several retinal proteins that elicit memory responses in uveitis patients also cause disease in animals suggests that similar mechanisms may occur in human disease. Likewise, this shows the heterogeneous nature of uveitis—both in mouse and man (Forrester et al. 2013). These various EAU models have allowed researchers and clinicians to establish that uveitis is a CD4+ T effector cell mediated disease with either Th1- or Th17-mediated responses (Caspi 2011). Also, the major cytokines identified in driving pro-inflammatory ocular disease processes include TNF-α, IFN-γ, IL-17, and IL-21, as well as many others (Forrester et al. 2013; Horai and Caspi 2011). The EAU model has served as an invaluable tool to evaluate novel immunotherapeutic modalities. Success in downregulating EAU in animals has often been predictive of clinical success of a given therapeutic agent in subsequent clinical trials.

Extrapolation from animal models has suggested that what triggers uveitis in humans could be an exposure to a retinal or antigenic mimic, combined with an infectious event that may provide inflammatory "danger" signals, creating conditions which lead to the development of ocular inflammation. Most EAU models of uveitis are dependent on activation of the innate immune system by use of adjuvants that activate a wide range of pathogen recognition receptors (PRRs). This raises the question of the underlying role of initial and/or persistent infection in immune-mediated uveitis (Forrester et al. 2013). Pathogenic mechanisms such as antigenic cross-reactivity and molecular mimicry may exist with pathogenic or commensal microorganisms. In addition, microbial components may act as "endogenous" adjuvants for provoking immune reactions to altered self-antigen, suggesting that there may be a common link in the pathogenesis of infectious and noninfectious human uveitis (Forrester et al. 2013).

6.5 Humanized Animal Models of Uveitis

Humanized animal models are a variant of induced EAU models. These have been established in the past decade as important tools for studying ocular inflammation including uveitis. Human uveitis is associated with specific HLA class I or class II types, and as a result the "humanized" EAU model has been developed based on the known susceptible human haplotypes (Pennesi et al. 2003; de Kozak et al. 2008; Mangalam et al. 2008).

HLA-A29 is associated with BC. A29-transgenic mice were developed in France using an HLA-A29 genetic construct cloned from the BC patient (de Kozak et al. 2008). According to the published study, this strain developed spontaneous uveitis late in life (12 months) with a frequency of approximately 80 %. This mouse model is important for studying BC and its correlation with HLA-29, but the late onset does make it a difficult experimental model to work with since the mice must age before the disease

manifests (de Kozak et al. 2008). Also, the target retinal antigen in this model is not known. But an even greater challenge is that the strain has become extinct and recent attempts to re-create the strain using the same genetic construct have yielded disappointing results (Mattapallil et al. 2012).

EAU can also be induced by immunization with the retinal antigen in HLA class II transgenic mice, in which the mouse class II molecules have been deleted and replaced with the human counterparts. HLA-DR3, HLA-DR4, HLA-DQ6, or HLA-DQ8 transgenic mice present and respond to antigenic epitopes that would be recognized by humans bearing these class II molecules (Mattapallil et al. 2008; Pennesi and Caspi 2002; Pennesi et al. 2003) Pennesi G. HLA-Tg mice develop severe uveitis when immunized with S-Ag and not only with IRBP. This is notable, because the original parental strains of mice from which these HLA-Tg strains have been derived are resistant to EAU induction with S-Ag. Since MHC class II molecules are involved in antigen recognition, the HLA class II Tg mice may help identify the critical regions of retinal antigens that are involved in human disease and may be important for finding antigen-specific therapies for human uveitis (Mattapallil et al. 2011). They will also be important tools to identify pathogenic epitopes and to study HLA-related cellular and genetic mechanisms of susceptibility.

6.6 Spontaneous Animal Models of Uveitis

6.6.1 Autoimmune Regulator (AIRE)-Deficient Mice

In most cases of uveitis, the disease forms spontaneously. Therefore, spontaneous models of uveitis are needed to mimic this disease feature in humans. Many tissue-specific antigens, including retinal antigens such as S-Ag and IRBP, are expressed ectopically in the thymus under the control of the transcription factor AIRE

(Forrester et al. 2013; Anderson et al. 2002). Self-antigens can be potential targets of autoimmunity and could be presented to the developing T cells. The T cells that express receptors having high affinity to self-antigens are eliminated or anergized, while those with low affinity to self-antigens are positively selected and exit the thymus into the periphery, where they control host defense. Targeted disruption of the *aire* genes causes multiorgan autoimmunity, including retinal antibodies, cellular responses to IRBP, and a spontaneous T cell mediated EAU-like uveitis (De Voss et al. 2006). IRBP appears to be the only antigen recognized by these mice as pathogenic. Mice that are *aire*-deficient develop spontaneous uveitis that increases in frequency and severity with age and is characterized by a mononuclear infiltrate into the eye and presence of serum autoantibodies specific to IRBP (Anderson et al. 2002; DeVoss et al. 2006). Aire-deficient mice are valuable for dissecting out the genetic, environmental, and stochastic processes in determining the target organ specificity of autoimmune destruction that can occur in the eye (Anderson et al. 2002). Although aire-deficient mice have not been well characterized concerning the development of spontaneous uveitis, they have potential to be an invaluable spontaneous mouse model of uveitis.

6.6.2 Transgenic Mice Expressing a T Cell Receptor Specific for IRBP

Recently, we have developed a new spontaneous mouse model directed at the native retinal antigen IRBP. This transgenic mouse (R161) on the EAU susceptible B10.RIII background expresses a TCR specific to the major uveitogenic epitope of IRBP (Horai et al. 2013). The IRBP TCR transgenic mice generate IRBP-specific effector cells in vivo and rapidly develop spontaneous uveitis, starting at weaning age. The disease pathology is similar to the EAU model (Chen et al. 2013a, b). Figure 6.2 shows the similarities in disease between an EAU mouse immunized with IRBP and an R161H mouse that developed spontaneous uveitis. The R161 mice have more chronic, progres-

sive disease compared with the EAU mice and are more likely to be physiologically relevant to the human disease state. This mouse model will provide insights on how autoreactive T cells specific to retina become primed and cause autoimmune uveitis. Three different lines of R161 transgenic mice were generated: R161High (R161H), R161Medium (R161M), and R161Low (R161L; Horai et al. 2013). Each of the lines has different levels of expression of the transgenic R161TCR and different proportions of IRBP-specific CD4$^+$ T cells in their periphery. Two of the lines R161H and R161M develop 100 % incidence by 2–3 months of age with moderate-to-severe disease. The R161L line develops trace disease with an incidence of less than 10 %. This mouse model and their IRBP-specific T cells derived from them will make it possible to study both the pathogenic and regulatory T cells specific to the native retinal antigen, avoiding the inherent self-antigen systems with adjuvants (Horai et al. 2013). More important, the three different lines will be valuable for assessing different severities of uveitis and may prove useful in preclinical studies and drug development (Chen et al. 2013a; Horai et al. 2013). The R161M and R161H lines will be useful in testing immunotherapies and drugs to reduce disease, while the R161L line will be useful in evaluating mechanisms involved in the transition from homeostasis to low-grade ocular inflammatory disease. In general, these spontaneous animal models can closely recapitulate some forms of human uveitis.

6.6.3 Expression of a Foreign Neo-Self-Antigen in the Eye: Hen Egg Lysozyme

Hen egg lysozyme (Hel) expressed in the lens under control of the αA-crystallin promoter allows development of a severe ocular inflammatory disease involving the retina and lens when single transgenic mice are crossed with the TCR Hel-Tg mouse (TCR-Hel; Zhang et al. 2003). Spontaneous disease also develops in mice that express Hel in the retina under control of the IRBP promoter when crossed with the TCR-

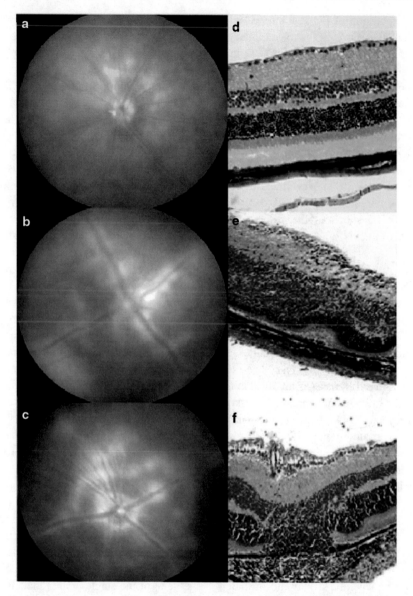

Fig. 6.2 Appearance of uveitis in an immunized mouse (EAU) with IRBP vs. a mouse with spontaneous uveitis (R161H) by fundus and histopathology. **a**, **c**, and **e** mouse fundus photographs. **b**, **d**, and **f** Hematoxylin and eosin-stained retinal cross sections. **a** and **b** shows a wild-type B10.RIII mouse retina with a normal intact retinal ar-chitecture. **c** and **d** shows the EAU model in a B10.RIII mouse that was immunized with IRBP. The fundus and retina displays significant retinal inflammation. **e** and **f** shows an R161H mouse that spontaneously developed uveitis. The fundus and retina show similar inflammation to the EAU mouse model represented in **c-d**

HEL-Tg mouse (Lambe et al. 2007). In both the models, 100 % of double transgenic mice develop the disease, with severe destruction and shrinkage of the eye globe, which can be seen in some cases of uncontrolled intraocular inflam-mation in humans (Forrester et al. 2013). In this, the Hel–Hel TCR double-transgenic models resemble the R161 IRBP TCR Tg mice dis-cussed above. It is unknown why a double-trans-genic situation with HEL expressed in the lens or retina in combination with the Hel-specific 3A9 TCR results in spontaneous uveitis, as does endogenous expression of IRBP combined with the R161 IRBP-specific TCR, but a combination of retinal β-gal with the β-gal-specific TCR does not (McPherson et al. 2009). However, in dealing with these models it must be kept in mind that in each case the mouse strains are different (B10. BR vs. B10.RIII vs. C57BL/6, respectively) and that expression of transgenes can be influenced by integration effects, resulting in an altered level of expression, which can cause variability in disease (Horai et al. 2013). Nevertheless, these spontaneous animal models are useful to address antigen-specific questions in the pathogenesis and regulation of uveitis.

6.6.4 TAM Knockout Mice

Ye and colleagues show that TAM knockout mice develop ocular inflammation and are more susceptible to immunization with IRBP peptides (Ye et al. 2011; Papotto et al. 2014). The TAM family of receptor tyrosine kinases is involved in the control of DC cytokine signaling, and knockout of the three receptors (TMO KO) of this family causes multiorgan autoimmune disease in mice (Lu and Lemke 2001). These mice display postnatal degeneration of the ocular photoreceptor layer and cellular infiltration by T lymphocytes and macrophages (Papotto et al. 2014). Moreover, IRBP-specific T lymphocytes are found in these mice, suggesting their susceptibility to the development of uveitis even when induced with low doses of IRBP. In this animal model, disease occurs spontaneously, and it may aid in understanding the contribution of dysregulated populations of APCs to the onset and maintenance of disease (Papotto et al. 2014).

6.7 Conclusions

Uveitis is a complex and diverse ocular disease. There are many similarities between animal models of uveitis and human clinical cases in terms of pathology and cellular mechanisms. Each animal

model described has unique features, which can aid in our understanding of different aspects of human uveitis. EAU is a well-established and commonly used animal model by vision researchers to study noninfectious uveitis. The model recapitulates similar clinical features found in human patients. Spontaneous genetic models and humanized models of uveitis are available, which model human disease more closely, especially chronic and relapsing forms of uveitis. Both the spontaneous and humanized animal models will also aid in testing the efficacy of new drugs and pharmaceuticals for ocular inflammation designed not only for systemic administration, but also for localized delivery into the eye.

Animal models of uveitis have greatly contributed to our understanding of the pathophysiology of noninfectious uveitis, but many questions remain such as the triggers of uveitis, recurrences, and susceptibility (Gasparin et al. 2012). Specifically, which microorganisms, including commensals and/or microbiota, may be involved in antigenic mimicry and which are the retinal-specific antigens that become activated to cause noninfectious human uveitis? Additionally, what are the cellular and molecular causes in the breakdown of ocular immune privilege and surveillance. Understanding how effector T cells enter the eye, breach the BRB, and mediate inflammatory responses will be crucial in identifying targets to block destructive Th1 and/or Th17 inflammatory processes from occurring in the eye. It is important to emphasize that although we have many different animal models of uveitis, none of the single animal models are ideal, nor can any truly mimic the human disease state (Bodaghi and Rao 2008). In fact, using combinations of different animal models is the more judicious approach in translational research. Testing of potential clinical therapies, designing immunological targets, and unveiling new pathogenic mechanisms should take place in more than one animal model of uveitis due to the heterogeneity of the disease. In summary, there are many animal models of uveitis, each with its own set of advantages and limitations. There is still a lot to uncover in the pathogenesis of uveitis concerning disease manifestation and resolution. There is no doubt that animal models will continue to be important tools for vision researchers to unravel the complexities of ocular inflammation.

6.7.1 Checklist of Core Messages (Bullet Points)

- Animal models of uveitis allow scientists and clinicians to study the basic mechanisms of inflammatory disease processes and serve as tools for the development of therapeutics and biologics.
- No single animal model reproduces the full complexity and heterogeneity of human disease. Each animal model of uveitis clearly has its own set of advantages and limitations. Therefore, the use of more than one model in translational research is deemed to be necessary.
- There are three major categories of animal models of uveitis: induced, spontaneous, and humanized models.
- Classical EAU induced by immunization with IRBP in CFA is a well-characterized and highly reproducible animal model. It has taught us that autoimmune uveitis is a T cell mediated disease. This model is typically acute and monophasic in its progression.
- Humanized models of uveitis include the HLA29 transgenic mice for class I dependence, and HLA DR3, DR4, DQ6, and DQ8 for class II dependence. These models will be important for studying pathogenic epitopes and HLA-related cellular and genetic mechanisms of susceptibility.
- Spontaneous EAU in IRBP TCR transgenic mice (R161) is a recently developed model, which will aid in clinical evaluation of drugs and therapeutics in three different lines with varying degrees of disease penetrance and severity. It avoids the use of adjuvants and other known immune stimulants to elicit disease manifestations.
- Spontaneous genetic models and humanized models of uveitis can more closely model

the human disease process, since these have chronic and relapsing phases of the disease, which are often observed in human uveitic diseases.

6.7.2 Compliance with Ethical Requirements

Jennifer L Kielczewski, Rachel R. Caspi and Robert Nussenblatt declare that they have no conflict of interest.

No human studies were carried out by the authors for this chapter. All institutional and national guidelines for the care and use of laboratory animals were followed.

References

Adamus G, Amundson D, Vainiene M, Ariail K, Machnicki M, Weinberg A, Offner H (1996) Myelin basic protein specific T-helper cells induce experimental anterior uveitis. J Neurosci Res 44(6):513–518

Agarwal RK, Caspi RR (2004) Rodent models of experimental autoimmune uveitis. Methods Mol Med 102:395–419

Agarwal RK, Silver PB, Caspi RR (2012) Rodent models of experimental autoimmune uveitis. Methods Mol Biol 900:443–469

Anderson MS, Venanzi ES, Klein L, Chen Z, Berzins SP, Turley SJ, von Boehmer H, Bronson R, Dierich A, Benoist C, Mathis D (2002) Projection of an immunological self shadow within the thymus by the aire protein. Science 298(5597):1395–1401

Bodaghi B, Rao N (2008) Relevance of animal models to human uveitis. Ophthalmic Res 40(3–4):200–202

Bora NS, Kim MC, Kabeer NH, Simpson SC, Tandhasetti MT, Cirrito TP, Kaplan AD, Kaplan HJ (1995) Experimental autoimmune anterior uveitis. Induction with melanin-associated antigen from the iris and ciliary body. Invest Ophthalmol Vis Sci 36(6):1056–1066

Broekhuyse RM, Kuhlmann ED, Winkens HJ, Van Vugt AH (1991) Experimental autoimmune anterior uveitis (EAAU), a new form of experimental uveitis. I. Induction by a detergent-insoluble, intrinsic protein fraction of the retinal pigment epithelium. Exp Eye Res 52(4):465–474

Broekhuyse RM, Kuhlmann ED, Winkens HJ (1992) Experimental autoimmune anterior uveitis (EAAU). II. Dose-dependent induction and adoptive transfer using a melanin-bound antigen of the retinal pigment epithelium. Exp Eye Res 55(3):401–411

Broekhuyse RM, Kuhlmann ED, Winkens HJ (1993) Experimental autoimmune anterior uveitis (EAAU).

III. Induction by immunization with purified uveal and skin melanins. Exp Eye Res 56(5):575–583

Caspi RR (2003) Experimental autoimmune uveoretinitis in the rat and mouse. Curr Protoc Immunol Chapter 15:Unit 15.16

Caspi RR (2006) Ocular autoimmunity: the price of privilege? Immunol Rev 213:23–35

Caspi R (2008a) Autoimmunity in the immune privileged eye: pathogenic and regulatory T cells. Immunol Res 42(1–3):41–50

Caspi RR (2008b) Immunopathology of the eye. Introduction. Semin Immunopathol 30(2):63–64

Caspi RR (2008c) Immunotherapy of autoimmunity and cancer: the penalty for success. Nat Revi Immunol 8(12):970–976

Caspi RR (2010) A look at autoimmunity and inflammation in the eye. J Clin Invest 120(9):3073–3083

Caspi RR (2011) Understanding autoimmune uveitis through animal models. The Friedenwald lecture. Invest Ophthalmol Vis Sci 52(3):1872–1879

Caspi RR, Roberge FG, Chan CC, Wiggert B, Chader GJ, Rozenszajn LA, Lando Z, Nussenblatt RB (1988) A new model of autoimmune disease. Experimental autoimmune uveoretinitis induced in mice with two different retinal antigens. J Immunol 140(5):1490–1495

Caspi RR, Silver PB, Luger D, Tang J, Cortes LM, Pennesi G, Mattapallil MJ, Chan CC (2008) Mouse models of experimental autoimmune uveitis. Ophthal Res 40(3–4):169–174

Chan CC, Hikita N, Dastgheib K, Whitcup SM, Gery I, Nussenblatt RB (1994) Experimental melanin-protein-induced uveitis in the Lewis rat. Immunopathologic processes. Ophthalmology 101(7):1275–1280

Chen J, Qian H, Horai R, Chan CC, Caspi RR (2013a) Use of optical coherence tomography and electroretinography to evaluate retinal pathology in a mouse model of autoimmune uveitis. PloS ONE 8(5):e63904

Chen J, Qian H, Horai R, Chan CC, Falick Y, Caspi RR (2013b) Comparative analysis of induced vs. spontaneous models of autoimmune uveitis targeting the interphotoreceptor retinoid binding protein. PloS ONE 8(8):e72161

Commodaro AG, Bueno V, Belfort R Jr, Rizzo LV (2011) Autoimmune uveitis: the associated proinflammatory molecules and the search for immunoregulation. Autoimmun Rev 10(4):205–209

Constantinescu CS, Lavi E (2000) Anterior uveitis in murine relapsing experimental autoimmune encephalomyelitis (EAE), a mouse model of multiple sclerosis (MS). Curr Eye Res 20(1):71–76

De Kozak Y, Camelo S, Pla M (2008) Pathological aspects of spontaneous uveitis and retinopathy in HLA-A29 transgenic mice and in animal models of retinal autoimmunity: relevance to human pathologies. Ophthal Res 40(3–4):175–180

De Voss J, Hou Y, Johannes K, Lu W, Liou GI, Rinn J, Chang H, Caspi RR, Fong L, Anderson MS (2006) Spontaneous autoimmunity prevented by thymic expression of a single self-antigen. J Exp Med 203(12):2727–2735

Forrester JV, Klaska IP, Yu T, Kuffova L (2013) Uveitis in mouse and man. Int Rev Immunol 32(1):76–96

Gasparin F, Takahashi BS, Scolari MR, Gasparin F, Pedral LS, Damico FM (2012) Experimental models of autoimmune inflammatory ocular diseases. Arq Bras Oftalmol 75(2):143–147

Gery I, Wiggert B, Redmond TM, Kuwabara T, Crawford MA, Vistica BP, Chader GJ (1986) Uveoretinitis and pinealitis induced by immunization with interphotoreceptor retinoid-binding protein. Invest Ophthalmol Vis Sci 27(8):1296–1300

Gregerson DS, Torseth JW, McPherson SW, Roberts JP, Shinohara T, Zack DJ (1999) Retinal expression of a neo-self antigen, beta-galactosidase, is not tolerogenic and creates a target for autoimmune uveoretinitis. J Immunol 163(2):1073–1080

Gritz DC, Wong IG (2004) Incidence and prevalence of uveitis in Northern California; the Northern California Epidemiology of Uveitis Study. Ophthalmology 111(3):491–500 (discussion 500)

Horai R, Caspi RR (2011) Cytokines in autoimmune uveitis. J Interferon Cytokine Res 31(10):733–744

Horai R, Silver PB, Chen J, Agarwal RK, Chong WP, Jittayasothorn Y, Mattapallil MJ, Nguyen S, Natarajan K, Villasmil R, Wang P, Karabekian Z, Lytton SD, Chan CC, Caspi RR (2013) Breakdown of immune privilege and spontaneous autoimmunity in mice expressing a transgenic T cell receptor specific for a retinal autoantigen. J Autoimmun 44:21–33

Jha P, Sohn JH, Xu Q, Nishihori H, Wang Y, Nishihori S, Manickam B, Kaplan HJ, Bora PS, Bora NS (2006) The complement system plays a critical role in the development of experimental autoimmune anterior uveitis. Invest Ophthalmol Vis Sci 47(3):1030–1038

Lai JC, Lobanoff MC, Fukushima A, Wawrousek EF, Chan CC, Whitcup SM, Gery I (1999) Uveitis induced by lymphocytes sensitized against a transgenically expressed lens protein. Invest Ophthalmol Vis Sci 40(11):2735–2739

Lambe T, Leung JC, Ferry H, Bouriez-Jones T, Makinen K, Crockford TL, Jiang HR, Nickerson JM, Peltonen L, Forrester JV, Cornall RJ (2007) Limited peripheral T cell anergy predisposes to retinal autoimmunity. J Immunol 178(7):4276–4283

Lee RW, Nicholson LB, Sen HN, Chan CC, Wei L, Nussenblatt RB, Dick AD (2014) Autoimmune and autoinflammatory mechanisms in uveitis. Semin Immunopathol 36:581–594

Levy RA, de Andrade FA, Foeldvari I (2011) Cutting-edge issues in autoimmune uveitis. Clin Rev Allergy Immunol 41(2):214–223

Li Q, Peng B, Whitcup SM, Jang SU, Chan CC (1995) Endotoxin induced uveitis in the mouse: susceptibility and genetic control. Exp Eye Res 61(5):629–632

Lu Q, Lemke G (2001) Homeostatic regulation of the immune system by receptor tyrosine kinases of the Tyro 3 family. Science 293(5528):306–311

Lyon F, Gale RP, Lightman S (2009) Recent developments in the treatment of uveitis: an update. Expert Opin Investig Drugs 18(5):609–616

Mangalam AK, Rajagopalan G, Taneja V, David CS (2008) HLA class II transgenic mice mimic human inflammatory diseases. Adv Immunol 97:65–147

Mattapallil MJ, Sahin A, Silver PB, Sun SH, Chan CC, Remmers EF, Hejtmancik JF, Caspi RR (2008) Common genetic determinants of uveitis shared with other autoimmune disorders. J Immunol 180(10):6751–6759

Mattapallil MJ, Silver PB, Mattapallil JJ, Horai R, Karabekian Z, McDowell JH, Chan CC, James EA, Kwok WW, Sen HN, Nussenblatt RB, David CS, Caspi RR (2011) Uveitis-associated epitopes of retinal antigens are pathogenic in the humanized mouse model of uveitis and identify autoaggressive T cells. J Immunol 187(4):1977–1985

Mattapallil MJ, Wawrousek EF, Chan CC, Zhao H, Roychoudhury J, Ferguson TA, Caspi RR (2012) The Rd8 mutation of the Crb1 gene is present in vendor lines of C57BL/6N mice and embryonic stem cells, and confounds ocular induced mutant phenotypes. Invest Ophthalmol Vis Sci 53(6):2921–2927

McPherson SW, Heuss ND, Gregerson DS (2009) Lymphopenia-induced proliferation is a potent activator for CD4+ T cell-mediated autoimmune disease in the retina. J Immunol 182(2):969–979

Nussenblatt RB (2002) Bench to bedside: new approaches to the immunotherapy of uveitic disease. Int RevImmunol 21(2–3):273–289

Papotto PH, Marengo EB, Sardinha LR, Goldberg AC, Rizzo LV (2014) Immunotherapeutic strategies in autoimmune uveitis. Autoimmun Rev 13:909–916

Pennesi G, Caspi RR (2002) Genetic control of susceptibility in clinical and experimental uveitis. IntRev Immunol 21(2–3):67–88

Pennesi G, Mattapallil MJ, Sun SH, Avichezer D, Silver PB, Karabekian Z, David CS, Hargrave PA, McDowell JH, Smith WC, Wiggert B, Donoso LA, Chan CC, Caspi RR (2003) A humanized model of experimental autoimmune uveitis in HLA class II transgenic mice. J Clin Invest 111(8):1171–1180

Rosenbaum JT, McDevitt HO, Guss RB, Egbert PR (1980) Endotoxin-induced uveitis in rats as a model for human disease. Nature 286(5773):611–613

Rothova A, Buitenhuis HJ, Meenken C, Brinkman CJ, Linssen A, Alberts C, Luyendijk L, Kijlstra A (1992) Uveitis and systemic disease. BrJ Ophthalmol 76(3):137–141

Rothova A, Suttorp-van Schulten MS, Frits Treffers W, Kijlstra A (1996) Causes and frequency of blindness in patients with intraocular inflammatory disease. Br J Ophthalmol 80(4):332–336

Silver PB, Chan CC, Wiggert B, Caspi RR (1999) The requirement for pertussis to induce EAU is strain-dependent: B10.RIII, but not B10.A mice, develop EAU and Th1 responses to IRBP without pertussis treatment. Invest Ophthalmol Vis Sci 40(12):2898–2905

Simpson E (2006) A historical perspective on immunological privilege. Immunol Rev 213:12–22

Smith JR, Rosenbaum JT, Williams KA (2008) Experimental melanin-induced uveitis: experimental model of human acute anterior uveitis. Ophthal Res 40(3–4):136–140

Srivastava A, Rajappa M, Kaur J (2010) Uveitis: Mechanisms and recent advances in therapy. Clin Chim Acta 411(17–18):1165–1171

Streilein JW (2003a) Ocular immune privilege: the eye takes a dim but practical view of immunity and inflammation. J Leukocyte Biol 74(2):179–185

Streilein JW (2003b) Ocular immune privilege: therapeutic opportunities from an experiment of nature. Nature Rev Immunol 3(11):879–889

Suttorp-Schulten MS, Rothova A (1996) The possible impact of uveitis in blindness: a literature survey. Br J Ophthalmol 80(9):844–848

Tang J, Zhu W, Silver PB, Su SB, Chan CC, Caspi RR (2007) Autoimmune uveitis elicited with antigen-pulsed dendritic cells has a distinct clinical signature and is driven by unique effector mechanisms: initial encounter with autoantigen defines disease phenotype. J Immunol 178(9):5578–5587

Terrada C, Fisson S, De Kozak Y, Kaddouri M, Lehoang P, Klatzmann D, Salomon BL, Bodaghi B (2006) Regulatory T cells control uveoretinitis induced by pathogenic Th1 cells reacting to a specific retinal neoantigen. J Immunol 176(12):7171–7179

Wildner G, Diedrichs-Mohring M, Thurau SR (2008) Rat models of autoimmune uveitis. Ophthal Res 40(3–4):141–144

Ye F, Li Q, Ke Y, Lu Q, Han L, Kaplan HJ, Shao H, Lu Q (2011) TAM receptor knockout mice are susceptible to retinal autoimmune induction. Invest Ophthalmol Vis Sci 52(7):4239–4246

Zeiss CJ (2013) Translational models of ocular disease. Vet Ophthalmol 16(Suppl 1):15–33

Zhang M, Vacchio MS, Vistica BP, Lesage S, Egwuagu CE, Yu CR, Gelderman MP, Kennedy MC, Wawrousek EF, Gery I (2003) T cell tolerance to a neo-self antigen expressed by thymic epithelial cells: the soluble form is more effective than the membrane-bound form. J Immunol 170(8):3954–3962

Commentary

Dr. Nussenblatt
e-mail: nussenblattr@nei.nih.gov
Laboratory of Immunology (UNITE Laboratory),
National Eye Institute,
Center for Human Immunology,
National Institutes of Health,
Bethesda, MD, USA

Kielczewski and Caspi provide the reader with a precise and well-written review of the animal models of ocular inflammatory disease. Really a collection of many clinical disorders, They are often sight threatening and require systemic immunosuppressive medication to treat the disease. Understanding the mechanisms leading to better therapy is clearly an important goal. It becomes even more important with evidence mounting that other disorders, such as age-related macular degeneration which accounts for much of the irreversible visual handicap in the Western World, may be immune mediated as well. Clearly many experimental manipulations cannot be performed in patients, and animal models, which are not perfect, provide the observer with the correct environment to observe and manipulate. Many potential mechanisms have been discerned using these models and there are still challenges. Correlating the observations from an animal model to the human can be a daunting task. In addition, our understanding of basic immune mechanisms will call for new models to be developed. With our better understanding of important additional elements such as the microbiome and epigenetics as well as the importance of noncoding DNA elements will present new challenges but also wonderful new opportunities as the scientific community moves forward to a new tomorrow.

Animal Models of Retinitis Pigmentosa (RP)

7

Bo Chang

7.1 Retinal Degeneration in Retinitis Pigmentosa Patients

Retinitis pigmentosa (RP) is the most common form of inherited retinal degeneration (RD) with a frequency of 1 in 3000–7000 individuals (Ferrari et al. 2011), with more than one million people affected worldwide. The term "retinitis" was first used in the 1850s because doctors thought the retina appeared inflamed. We now know that most RP is not caused by infection, but is generally an inherited condition. The word "pigmentosa" refers to the pigment deposits seen during the examination of the fundus. Typical symptoms include night blindness followed by decreasing visual fields, leading to tunnel vision and eventually legal blindness or, in many cases, complete blindness. On the cellular level, this correlates with a predominantly dysfunctional rod photoreceptor system. In later stages, the disease may further affect cone photoreceptors, eventually causing complete blindness. Findings on retinal examination include "bone spicule" pigmentary deposits, retinal vessel attenuation, abnormal, diminished or absence of a- and b-waves in electroretinogram (ERG), and reduced visual field (Daiger et al. 2013). RP can be inherited in an autosomal dominant, autosomal recessive, or X-linked manner. While usually limited to the eye, RP may also occur as part of a syndrome as in the Usher syndrome and Bardet–Biedl syndrome. Over 50 genes have been associated with RP and nearly 3100 mutations have been reported in these genes (Daiger et al. 2013). Among the autosomal recessive RP (arRP) genes identified thus far, mutations within *PDE6A* (coding for phosphodiesterase 6A) and *PDE6B* (coding for phosphodiesterase 6B), the genes associated with the PDE6 complex that are essential for photoreceptor function and maintenance are, in fact, the second most common identifiable cause of arRP. About 36,000 cases of simplex and familial RP worldwide are due to defects in the PDE6 complex, estimated to account for approximately 8 % of all diagnosed arRP (Ferrari et al. 2011; Daiger et al. 2013). Mutations in the gene encoding the beta subunit of rod phosphodiesterase (PDE6B) are the most common identified cause of arRP, accounting for approximately 4 % of cases in North America (McLaughlin et al. 1995).

7.2 RD in Animal Models

Models of RD have been discovered or generated in mice, rats, rabbits, pigs, zebrafish, and nonhuman primates. Mouse models offer the advantages of low cost, disease progression on a relatively quick time scale, and the ability to perform genetic manipulations. The ability to target and alter a specific gene(s) is an important and necessary tool to produce mouse models

B. Chang (✉)
The Jackson Laboratory, 600 Main Street, Bar Harbor, ME 04609, USA
e-mail: bo.chang@jax.org

© Springer International Publishing Switzerland 2016
C.-C. Chan (ed.), *Animal Models of Ophthalmic Diseases*, Essentials in Ophthalmology,

with mutations in genes of choice. Inducing mutations in genes of choice (known as knockout or transgenic) is termed as "reverse genetics," while "forward genetics" approaches initiate as spontaneous/induced mutations that are discovered as a result of the overt phenotypes, and the underlying mutation is subsequently identified. Through the "forward" and "reverse" genetic approaches, mouse models in >100 genes that underlie human retinal diseases have been studied (Chang et al. 2007a; Samardzija et al. 2010; Won et al. 2011).

Mouse models of RD have been known for many years. The first RD model, discovered by Dr Clyde E. Keeler more than 90 years ago (Keeler 1924, 1966), was identified as a mutation in *Pde6b* (formerly *rd1*, *rd*, identical with Keeler rodless retina, *r*) (Pittler et al. 1993). Frequently, mutated genes are identified first in the mouse. For example, the rod phosphodiesterase beta-subunit (PDE6B) and peripherin/RDS were first found to be the cause of RD in mice before their connection to human diseases was demonstrated. Discovery of the mouse retinal degeneration 1 (*rd1*) mutation in the gene for the beta-subunit of cGMP-phosphodiesterase (*Pde6b*) (Bowes et al. 1990) led to the identification of mutations in the human homolog (*PDE6B*) in similar human disorders (McLaughlin et al. 1993), and discovery of the mutation *Rds* (retinal degeneration slow) in the peripherin 2 (*Prph2*) (Travis et al. 1989) paved the way for the identification of defects in the RDS/peripherin gene in patients with autosomal dominant RP and various cone, cone–rod, and macular dystrophies (Keen and Inglehearn 1996). Identification of the tubby (*tub*) mouse gene family (Noben-Trauth et al. 1996) led to the finding of tubby-like protein 1 (*TULP1*) alterations in individuals affected with arRP (Hagstrom et al. 1998; Banerjee et al. 1998). Similarly, the shaker 1 (*sh1*) mutation in the myosin 7A (*Myo7a*) (Gibson et al. 1995) mouse gene led to identification of mutations in *MYO7A* in patients with Usher's syndrome type 1b (Weil et al. 1995). Additionally, it was the studies on mouse that led to the identification of apoptosis as an important mechanism of photoreceptor cell death in RD (Chang et al. 1993; Portera-Cailliau et al. 1994).

7.3 Using the *rd10* Mouse Model for studies of RP

The retinal degeneration 10 (*rd10*) model was discovered more than 10 years ago (Chang et al. 2002). Mice homozygous for the *rd10* mutation show grainy retinas with sclerotic retinal vessels by indirect ophthalmoscopy at 4 weeks of age. Histology at 3 weeks of age shows photoreceptor degeneration that progress as mice age. Electroretinograms of *rd10/rd10* mice are never normal. The maximal response occurs at 3 weeks of age and is not detectable at 2 months of age. Genetic analysis showed that this disorder was caused by an autosomal recessive mutation that mapped to mouse Chr 5. Sequence analysis showed a missense mutation in exon 13 of the beta subunit of the rod phosphodiesterase gene. Therefore, the gene symbol for the *rd10* mutation is *Pde6b^{rd10}* (Chang et al. 2002). In contrast to *rd1*, the onset of RD in the *rd10* mouse is later and can be delayed by rearing in darkness. Also, while the PDE6B protein and activity in *rd1* mutant mice are undetectable, they are detected early, at P10, in *rd10* mouse retinas with western blotting and immunostaining. However, the protein level is significantly reduced compared with that of age-matched wild-type controls (Chang et al. 2007b). The residual activity of PDE6B in *rd10* retinas may prevent a toxic increase in cGMP early in the mouse's life, thus delaying the onset of photoreceptor cell death, and extending the window for therapeutic intervention. Many missense, pathogenic human mutations in *PDE6B* leading to arRP are located within the catalytic domain (McLaughlin et al. 1995), potentially resulting in partial loss of function and reduced PDE6B enzymatic activity, as seen in the *rd10* allele. Thus, the *rd10* represents a better mouse model than *rd1* for developing strategies for treating human patients with recessive RP (Han et al. 2013).

7.3.1 Methodology Used to Characterize the *rd10* Mouse Model

Many of the clinical and functional methods used in routine eye examinations in human patients have been adapted to examine the small eyes of the mouse. Additionally, invasive methods that cannot be used in human patients are available and being developed to further our understanding of progression and pathology of retinal diseases. In the following paragraphs, we describe the methods that have been used to characterize the *Pde6b^{rd10}* model (herein referred to as *rd10*), which is proving to be an invaluable model for understanding RP and for testing different therapeutic approaches.

Indirect ophthalmoscopy, a noninvasive clinical method, which allows for the examination of the fundus, is a primary screen used in mice. However, due to the small size of the mouse eye, a 60–90 diopter lens is used to magnify the fundus. Using this method, retinal vessel attenuation is observed at 3 weeks of age and "bone spicule" pigmentary deposits at 2 months of age in *rd10* mutants (Chang et al. 2007b). Spectral domain optical coherence tomography (sd-OCT) scans are another noninvasive clinical technique that can be a powerful tool to monitor the course of RD with respect to morphometric changes. In *rd10* mice, retinal thickness declines up to 9 weeks of age; the loss of thickness is predominantly due to the thinning of the outer nuclear layer (ONL). Subsequently, the ONL is no longer visible by OCT, while the thickness of inner retinal layers remains nearly constant (Rösch et al. 2014). A previously unreported phenotype was observed using OCT, a separation of the retina from the retinal pigmented epithelium (RPE) was reported in some *rd10* mutants at P27, but by P64, almost all *rd10* mice demonstrated some degree of retinal separation. The retinal separation was never seen in *rd1* mice, which bears a mutation in the same *Pde6b* gene (Pennesi et al. 2012). In contrast, retinal separation was only observed in two of six mice examined by OCT, one 9-week-old and another 24-week-old *rd10* mouse by Rösch

and coworkers (Rösch et al. 2014), suggesting potential segregating genetic background modifiers or environmental interactions with the mutant allele.

The sd-OCT has also been used to evaluate a therapeutic endpoint in *rd10* mice. The ability to compare relative retinal thicknesses noninvasively in treated versus untreated eyes of *rd10* mice provides a valuable tool in which to score the effects of therapeutic intervention. The ability to monitor therapeutic outcome (retinal thickness) longitudinally without terminating the experimental animals greatly enhances the ability to carry out long-term rescue studies (in the *rd10* model as well as in others) and dramatically decreases the number of animals required for such studies (Pang et al. 2011). Finally, blood flow (BF) and anatomical magnetic resonance imaging (MRI) can also be used as a clinical method of assessing retinal alterations. In *rd10* mice, retinal BF decreased progressively over time while choroidal BF was unchanged. Similar to observations using OCT, by anatomical MRI, the *rd10* retina became progressively thinner at later time points compared with that of age-matched controls (Muir et al. 2013).

While clinical methods allow for noninvasive approaches to examine potential morphometric alterations as a consequence of disease, it is also important to be able to detect functional abnormalities that may or may not accompany morphological defects. Electroretinography is a functional test that assesses the response of photoreceptors and secondary neurons to light stimulation. The *rd10* mouse ERG amplitude is smaller at all flash intensities at 18 days of age compared to wild-type responses, and there is a progressive loss of cone function initiating at P30. However, the rate of cone function loss is not as fast as rod function loss in *rd10* mutants. At P18, the *rd10* dark-adapted response was 30% of the C57BL/6 response while the light-adapted response was 50% of the wild-type response. Similarly, at P30, the *rd10* dark-adapted b-wave was only 10% of controls while the light-adapted response was 30% (Chang et al. 2007b; Jae et al. 2013; Rösch et al. 2014). The optomotor test is a noninvasive technique, widely used for the evaluation of visual function in rodents. It is based on the optokinetic reflex that causes the animals to track a drifting stripe pattern with eye and head movements. Optometry demonstrated that the *rd10* mutants show a progressive decline in visual performance (Benkner et al. 2013; Thomas et al. 2010). The optomotor test, which is used to test the efficacy of AAV8 (Y733F)-smCBA-PDE β for rescue of both photopic and scotopic vision in *rd10* mice, demonstrated that visual performance was significantly better in treated *rd10* eyes compared with untreated *rd10* eyes (Pang et al. 2011).

Animal models allow for the direct examination of changes by invasive methods that cannot be reasonably done in human patients. Standard histology of tissue specimens has been commonly used to assess changes in particular cell types in the retina. For example, histological examination revealed progressive retinal ONL degeneration in *rd10* mice that initiates in the central retina at P16 and spreads to the peripheral retina by P20. By P25, only two to three layers of nuclei are left, and by P45, only a single layer of photoreceptors is left. Complete absence of photoreceptor nuclei is observed by P60. However, the cell number within the inner nuclear and ganglion cell layers does not appear to be affected (Chang et al. 2007b; Gargini et al. 2007). While histology has been considered the gold standard for assessing and quantifying the rate of RD in animal models, as well as for judging the efficacy of potential treatments, many disadvantages are associated with histologic processing. Because of the inherent need to sacrifice animals to obtain histologic data, an increase in the number of sequential time points in a long-term study greatly increases the total number of animals needed to conduct histologic analyses. Additionally, gathering histologic samples makes it impossible to perform such studies on the same animal over time. OCT and MRI may potentially be used to replace or at least significantly reduce the number of specimens examined by histology, as histologic results have been correlated and confirmed by OCT (Rösch et al. 2014; Pennesi et al. 2012; Pang et al. 2011) and by MRI (Muir et al. 2013) in *rd10* mutants. There are situations, however, in which direct

examination of the tissue may provide insights into the disease phenotypes observed. A number of investigators have used RPE whole mounts to examine the effects of the *Pde6b^{rd10}* mutation on RPE cells, which functionally and structurally support the overlying photoreceptor cells. In normal retinas, the RPE morphology resembles a regular hexagonal array of cells of uniform size. In the *rd10* retina, a clear disruption of RPE cell size and shape is observed (Chrenek et al. 2012; Jiang et al. 2013, 2014). This RPE abnormality might be the cause for the retinal separation (retinal detachment) (Jae et al. 2013; Rösch et al. 2014) observed in the *rd10* mouse. Further study is needed to determine the cause of the retinal detachment in *rd10* mutants that is not observed in *rd1* mutants.

Immunohistochemistry (IHC) allows for the localization of antigens or proteins in tissue sections by the use of labeled antibodies as specific reagents through antigen–antibody interactions that are visualized by a marker such as fluorescent dye, enzyme, or colloidal gold. It has been widely used in *rd10* mice. By using a PDE6b polyclonal antibody, staining was found in the inner and outer segments of photoreceptors in cryosections of P22 wild-type and *rd10* retina, but not in age-matched *rd1* retina cryosections, indicating the delayed photoreceptor degeneration in *rd10* mutants compared with *rd1* mutants. But the level of PDE6B expression in *rd10* mutants was significantly reduced compared with that of wild-type control (Chang et al. 2007b). Other groups also have used IHC to evaluate the progression of neuronal remodeling of second- and third-order retinal cells and their synaptic terminals in retinas from *rd10* mice at varying stages of degeneration ranging from P30 to postnatal month 9.5. Significant remodeling is observed in *rd10* mutants relative to wild-type control mice, especially at later stages of disease progression, as assessed by examination of cell- and synapse-specific markers (Phillips et al. 2010, 2011; Mazzoni et al. 2008; Barhoum et al. 2008).

Because of the milder disease phenotype in *rd10* mice, in which photoreceptor loss progresses more slowly than in *rd1* mice, examination of the effects of photoreceptor loss on the remaining retinal structures can be done. For example, Menzler et al. (2014) examined the retinal ganglion cell (RGC) structure and function in *rd10* mice. RGCs are the only retinal neurons that project to the brain and their normal structure and function are very important for vision. Using a single-cell approach, the investigators showed that all types of RGCs in *rd10* mice retain their dendritic architecture and overall viability, even after complete photoreceptor degeneration and regressive remodeling of the retina's second-order neurons up to 9 months of age (Menzler et al. 2014). Rhythmic RGC activity in healthy retinas was detected upon partial photoreceptor bleaching using an extracellular high-density multi-transistor array. The mean fundamental spiking frequency in bleached retinas was 4.3 Hz; close to the RGC rhythm detected in blind *rd10* mouse retinas (6.5 Hz) (Biswas et al. 2014). Other studies showed that the rhythmic RGC activity in *rd10* mouse retinas was around 3–7 Hz, and in the *rd1* mouse retinas was around 10–16 Hz (Goo et al. 2011a, b). These results suggest that in *rd10* retinas, the retinal circuitry may be less affected by photoreceptor loss. Further evidence is presented in a study on light-evoked RGC activity (Stasheff et al. 2011), where the light response properties of wild-type mice and *rd10* mice RGCs at P28 appear very similar, which is a strong indicator that basic circuitry evolved normally in *rd10*. The *rd1* retina, on the other hand, seems to possess a more defective circuitry.

The *rd10* mouse model has also been used to validate new in vitro methods, which in turn have provided additional characterization of the model. An example is the use of a laser micrometer to make noncontact measurements of linear external dimensions of the mouse eye. The resolution of the laser micrometer is less than 0.77 μm, and it provides accurate and precise noncontact measurements of eye dimensions on three axes: axial length (anterior–posterior (A–P) axis) and equatorial diameter (superior–inferior (S–I) and nasal–temporal (N–T) axes). External dimensions of the eye strongly correlated with eye weight. The *rd10* mice were found to have a slightly smaller eye size and weight when compared with that of WT C57BL/6J mice, which might cause the hyperopia observed in *rd10* mutant mice (Wisard et al. 2010). The *rd10* mice had significantly greater hyperopia relative to the WT controls throughout normal development; however, axial length became significantly shorter in *rd10* mutants relative to WT mice starting at 7 weeks of age. Interestingly, while RD models with low basal levels of DOPAC (3,4-dihydroxyphenylacetic acid) in general showed an increased susceptibility to form deprivation myopia; refractive development under normal visual conditions was disrupted toward greater hyperopia from 4 to 12 weeks of age in the *rd10* mice, despite significantly lower DOPAC levels. These results indicate that photoreceptor degeneration may alter dopamine metabolism, leading to increased susceptibility to myopia with an environmental visual challenge (Park et al. 2013).

7.3.2   Studies Focused on the Underlying Cellular and Molecular Changes in the *rd10* Mouse Model to Provide Insights About RP

A number of approaches have been taken to elucidate the cellular responses that result from the *rd10* mutation, which include both global- and pathway-specific approaches. A global transcriptome study carried out by Uren and coworkers (Uren et al. 2014) found a total of 1079 genes with significant changes in mRNA levels. Of those, 385 showed a relative increase in expression, while 694 showed a relative decrease in expression in *rd10* retinas compared with that of controls. Genes with relative decreased expression were mainly rod-specific genes, and their extreme decrease is expected, driven by the loss of rod photoreceptors. Genes with relative increased expression were predominantly Müller-specific genes; the proportion of which was significantly higher than expected under the null hypothesis that increased transcript expression would not favor any particular cell type. This suggests either increased numbers or modified activity of Müller glial cells. In addition to the prominent changes observed in photoreceptor and Müller cells, a relative decrease in expression of a few amacrine and bipolar cell-specific genes and increase in amacrine, ganglion, and bipolar-specific genes were also observed. This indicates transcriptional changes in cells other than rods and Müller glia (Uren et al. 2014). This study implicates remodeling of the inner retina and possible Müller cell dedifferentiation in the *rd10* mouse retina.

Pathway-specific transcriptome analysis of survival pathways showed that the *rd10* retina appears to attempt to protect photoreceptors from death and provides an excellent model to study the molecular mechanisms of neuroprotection and cell death. The degenerating *rd10* retina rapidly induces expression of transcriptional regulators *Atf3* (transcription factor 3) and *Cebpd* (CCAAT/enhancer-binding protein delta). Induction of *Atf3* was transient and only lasted for several days at the beginning of degeneration, whereas levels of *Cebpd* remained elevated throughout the period of photoreceptor loss. Several protective genes such as *Lif* (leukemia inhibitory factor), *Edn2* (endothelin-2), and *Fgf2* (fibroblast growth factor-2), which are implicated in a potent endogenous survival pathway, and *Mt1* (metallothionein-1) and *Mt2,* were strongly upregulated in the *rd10* retina. In addition, increased expression of *Casp1* (caspase-1) and *Il1b* (interleukin-1β) suggested an inflammatory response (Samardzija et al. 2012). Other studies also demonstrated that when retinal cells are stressed there is an initial struggle to survive, mediated through inhibition of PP2A (protein phosphatase 2A) and subsequent upregulation of survival pathways in the *rd10* mouse (Finnegan et al. 2010).

While in general, rod photoreceptor cell death occurs through apoptosis, necrosis, a form of cell injury that results in autolysis of living tissue, can also cause premature cell death. Ultrastructural analysis of *rd10* mouse retinas revealed that a substantial fraction of dying cones exhibited necrotic morphology. RIP1 and RIP3, two members of the receptor-interacting protein (RIP) kinase family of proteins, have been identified as critical mediators of programmed necrosis, and interestingly, *Rip3* deficiency is able to mediate partial rescue of cells in *rd10* mutants. Additionally, pharmacologic treatment with a RIP kinase inhibitor attenuated histological and functional deficits of cones in *rd10* mice (Murakami et al. 2012b).

Recently, there has been a great deal of interest in the role of microglia homeostasis in retinal diseases (Karlstetter et al. 2010). Notwithstanding, activated microglia have been reported to be an important contributor to the overall apoptosis of photoreceptors in *rd10* retinas. The suppression of microglia activation by minocycline or SC-560, a cyclooxygenase-1 (COX-1)-specific inhibitor, reduced microglia-mediated photoreceptor death, and significantly improved retinal structure and function and visual behavior in *rd10* mice. Their neuroprotective effects are thought to occur through both anti-inflammatory and antiapoptotic mechanisms (Peng et al. 2014). The activated *Cx3cr1* (chemokine (C-X3-C motif) receptor 1, deficient of microglia, had an additional neurotoxic effect on photoreceptor survival in *rd10* retinas, suggesting that the *Cx3cl1* (chemokine (C-X3-C motif) ligand 1)/ Cx3cr1 signaling pathway might protect against microglia neurotoxicity (Peng et al. 2014, Zieger et al. 2014). In comparison with wild-type controls, retinas of *rd10* mice have significantly lower levels of total CX3CL1 (fractalkine) protein (from P10 onwards) and lower CX3CL1 mRNA levels (from P14), even before the onset of primary rod degeneration (Zieger et al. 2014). The *rd10* mouse retina exhibits high levels of photoreceptor cell death and reactive Müller gliosis early in the degenerative process. In explant cultures, both degenerative processes were abrogated by insulin-like growth factor 1 (IGF-1) treatment. Moreover, the beneficial effect of IGF-I was diminished by microglial depletion using clodronate-containing liposomes. Interestingly, in the absence of IGF-I, microglial depletion partially prevented the cell death without affecting Müller gliosis. These findings strongly suggest a role for microglia–Müller glia crosstalk in neuroprotection, and the beneficial neuroprotective effects may be achieved through strategies that modulate microglial cell responses (Arroba et al. 2011, 2014). Neuroinflammation involving CC chemokine (or β-chemokine) proteins such as monocyte chemoattractant protein-1 (MCP-1) and its receptor CCR2 (chemokine (C-C motif) receptor 2) have been studied in the *rd10* retina and results suggest that the MCP-1/CCR2 system plays a role in RD (Guo et al. 2012). The number of microglia in the retinas of the *Ccr2−/−; rd10/rd10* double mutant mice was significantly lower than that in the retinas of mutant mice bearing the *rd10/rd10* mutation alone. The average number of cells in the ONL and the ERG amplitude were significantly higher in the *Ccr2−/−; rd10/rd10* double-mutant mice than in the *rd10/rd10* single-mutant mice (Guo et al. 2012). Finally, Yoshida and coworkers have shown that the sustained chronic inflammatory reaction contributes to the pathogenesis of RD in *rd10* mice and antioxidant treatment with *N*-acetylcysteine (NAC) prevented the photoreceptor cell death along with suppression of inflammatory factors and microglial activation (Yoshida et al. 2013). This suggests that an intervention for ocular inflammatory reaction using antioxidants may be a potential treatment for patients with RP.

The latter study implicates oxidative stress in the disease pathology observed in *rd10* mutants. Oxidative stress reflects an imbalance between the systemic manifestation of reactive oxygen species and a biological system's ability to readily detoxify the reactive intermediates or to repair the resulting damage. DNA oxidation was increased in an early phase of RD in the *rd10* mouse. Transgenic overexpression of human MutT homolog-1(hMTH1) in the *rd10* mouse attenuated accumulation of oxidative DNA damage, prevented the subsequent poly(ADP-ribose) polymerase (PARP) activation, and delayed photoreceptor cell death (Murakami et al. 2012). The metallocomplex zinc-desferrioxamine (Zn/DFO) may ameliorate the oxidative stress by chelation of labile iron in combination with release of zinc. Zn/DFO-treated *rd10* mice showed significantly higher electroretinographic responses at 3 and 4.5 weeks of age compared with saline-injected controls. Corresponding retinal (photoreceptor)

structural rescue was observed by quantitative histological and immunohistochemical techniques (Obolensky et al. 2011). Coexpression of superoxide dismutase 1 (SOD1) with glutathione peroxidase 4 (Gpx4), which like SOD1 is localized in the cytoplasm, but not with catalase targeted to the mitochondria, reduced oxidative stress in the retina and significantly slowed the loss of cone cell function in *rd10* mice (Usui et al. 2011). The benzopyran BP (3,4-dihydro-6-hydroxy-7-methoxy-2,2-dimethyl-1(2H)-benzopyran) is a free radical scavenger that is structurally similar to alpha-tocopherol and has provided neuroprotection in a number of disease models where oxidative stress is a causative factor. A novel derivative of BP with improved lipid solubility has been designated BP3. Systemic dosing with BP3 on alternate days between postnatal days 18 and 25 preserved rod photoreceptor numbers and cone photoreceptor morphology in *rd10* mice (O'Driscoll et al. 2011). Orally administered NAC reduced cone cell death and preserved cone function by reducing oxidative stress in the *rd10* and *rd1* mice. In *rd10* mice, supplementation of drinking water with NAC promoted partial maintenance of cone structure and function for at least 6 months. Topical application of NAC to the cornea of *rd10* mice also reduced superoxide radicals in the retina and promoted survival and functioning of cones (Lee et al. 2011). Two constituents of bile, bilirubin and tauroursodeoxycholic acid (TUDCA), have antioxidant activity. In *rd10* mice, intraperitoneal injections of 5 or 50 mg/kg of bilirubin or 500 mg/kg of TUDCA every 3 days starting at P6 caused significant preservation of cone cell number and cone function at P50. Rods were not protected at P50, but both bilirubin and TUDCA provided modest preservation of ONL thickness and rod function at P30 (Oveson et al. 2011). Other studies showed that TUDCA treatment preserved rod and cone function and greatly preserved overall photoreceptor numbers (Drack et al. 2012; Phillips et al. 2008; Boatright et al. 2006). *Para*-aminobenzoic acid (PABA) is a cyclic amino acid, which may act to decrease lipid peroxidation and oxidative stress. PABA (50 mg/kg) was administered intraperitoneally six times per week in *rd10* mice from P3.

PABA treatment may protect retinal function and attenuate the course of RD in *rd10* mice. Biochemical parameters indicate a lower degree of oxidative injury in PABA-treated retinas (Galbinur et al. 2009). Coexpression of superoxide dismutase 2 (SOD2) and catalase, but neither alone, significantly reduced oxidative damage in the retinas of P50 *rd10* mice as measured by protein carbonyl content. Cone density was significantly greater in P50 *rd10* mice with coexpression of SOD2 and catalase together than *rd10* mice that expressed SOD2 or catalase alone, or expressed neither (Usui et al. 2009). Systemic treatment of *rd10/rd10* mice with a mixture of antioxidants, including α-tocopherol, ascorbic acid, MnT-BAP, and α-lipoic acid between P18 and P35, showed preservation of cone function as shown by a significant increase in photopic ERG b-wave amplitudes, and surprisingly showed temporary preservation of scotopic a-wave amplitudes and prolonged rod survival (Komeima et al. 2007). Members of the insulin family are well-characterized neuroprotective molecules and expression of human proinsulin in *rd10* mice attenuates RD, as assessed by the maintenance of the ERG and the histologic preservation of photoreceptors. Systemic proinsulin was able to reach retinal tissue, delay apoptotic death of photoreceptors, and decrease oxidative damage (Corrochano et al. 2008). Iron is suggested as a potential source of oxidative radicals and it is associated with the RD processes in the *rd10* mouse by generating oxidative injury (Deleon et al. 2009; Picard et al. 2010; Jeanny et al. 2013). Enhanced survival of cones and rods in the retina has been demonstrated through overexpression or systemic administration of human transferrin (hTf) in *rd10* mice (Picard et al. 2010; Jeanny et al. 2013).

7.3.3 Therapeutic Interventions Tested in the *rd10* Mouse Model to Improve Photoreceptor Survival and/or Visual Response

In addition to those studies described above in which anti-inflammatory or antioxidant treatment was used to delineate pathological pathways induced by the *rd10* mutation, a host of investigators have used the *rd10* mutant, with its milder disease progression, to test the efficacy of therapeutic interventions at both the global and pathway levels. For example, environmental enrichment (EE), an experimental manipulation based on exposure to enhanced motor, sensory, and social stimulation, when started at birth, has been shown to exert clear beneficial effects by slowing vision loss. Likewise, early exposure to EE efficiently delayed photoreceptor degeneration in *rd10* mutants, extending the time window of cone viability and cone-mediated vision well beyond the phase of maximum rod death (Barone et al. 2012). Cone counting in retinal whole mounts showed that *rd10* EE mice at 1 year had almost three times as many surviving cones ($34,000 \pm 4000$) as the standard (ST) laboratory condition control mice ($12,700 \pm 1800$, t test, $p = 0.003$). Accordingly, the EE-treated *rd10* mutants at 1 year of age were still capable of performing the visual water task under photopic conditions, showing a residual visual acuity of 0.138 ± 0 cycles/degree. This ability was virtually absent in the *rd10* ST age-matched mice (0.063 ± 0.014, t test, $p = 0.029$) (Barone et al. 2014).

Neurotrophic factors have also been suggested as a potential global approach to promote photoreceptor survival. Long-term expression of glial cell line-derived neurotrophic factor (GDNF) in the *rd10* mouse had significant increases in ONL thickness and mean photopic and scotopic ERG b-wave amplitudes compared with that of the untreated *rd10* mouse (Ohnaka et al. 2012). Erythropoietin (Epo), a neurotrophic molecule discovered in the retina, administered systemically leads to morphological photoreceptor protection in the light-damage and RDS/peripherin (Prph2) models of RD. In contrast, no photoreceptor protection was observed following Epo gene transfer in the *rd10* model (Rex et al. 2004). Valproic acid (VPA) is an anticonvulsant and mood-stabilizing drug and it has the potential to increase the expression of neurotrophic factor (NF) genes. A single systemic dose of VPA can change retinal NF such as brain-derived neurotrophic factor

(BDNF), GDNF, ciliary neurotrophic factor (CNTF), fibroblast growth factor-2 (FGF2/ bFGF), and rod-specific gene expression in the immature retina. Daily VPA treatment from P17 to P28 can also alter gene expression in the mature neural retina. While daily treatment with VPA could significantly reduce photoreceptor loss in the *rd1* model, VPA treatment slightly accelerated photoreceptor loss in the *rd10* model (Mitton et al. 2014).

A number of antiapoptotic compounds have also been tested in *rd10* mice, with mixed results. Norgestrel is a synthetic progestin commonly used in hormonal contraception and it has a novel antiapoptotic role in diseased mouse retinas in vivo. Apoptosis plays a crucial role in developing and maintaining the health of the body by eliminating old cells, unnecessary cells, and unhealthy cells. It does so in a programmed sequence of events that lead to the elimination of cells without releasing harmful substances into the surrounding area. Norgestrel treatment preserves photoreceptor number, morphology, and function in *rd10* mice. Norgestrel improves ERG responses in *rd10* mice, indicating partial rescue of retinal function. Its neuroprotective mechanism of action is likely to involve basic fibroblast growth factor (bFGF) and extracellular signal-regulated kinases 1/2 (Erk1/2) as increases in their expression and activation have been observed (Doonan et al. 2011). Another potential neuroprotective treatment, polysaccharides of wolfberry, is long known to possess primary beneficial properties in the eyes. These polysaccharides provide long-term morphological and functional preservation of photoreceptors and improved visual behaviors in *rd10* mice. Moreover, the neuroprotective effects of the polysaccharides appear to occur through antioxidant, anti-inflammatory, and antiapoptotic mechanisms. Furthermore, the polysaccharides modulate inflammation and apoptosis partly through inhibition of NF-κB and HIF-1α expression, respectively (Wang et al. 2014). Photoreceptor degeneration is associated with calcium overload and calpain activation, both of which are observed before the appearance of signs of photoreceptor degeneration in *rd10* mice. In-

terestingly, photoreceptor death is increased by the induction of autophagy with rapamycin and inhibited by calpain and cathepsin inhibitors, both ex vivo and in vivo (Rodríguez-Muela et al. 2014). Another molecule, cell division cycle 42 homolog (*Saccharomyces cerevisiae*; CDC42), has been linked to cellular processes including movement, development, and apoptosis. CDC42 accumulated in the perinuclear region of photoreceptors in *rd10* mice during RD, but knockdown of CDC42 did not affect retinal morphology or function in the adult mice, and did not influence photoreceptor apoptosis or molecular signaling (Heynen et al. 2011). Finally, Sirtuin 1 (SIRT1), a nicotinamide adenine dinucleotide (NAD)-dependent deacetylase, influences a wide range of cellular processes like aging, transcription, apoptosis, and stress resistance. The antiapoptotic, neuroprotective role of *Sirt1* in the mouse retina is based on the involvement of *Sirt1* in DNA double-strand break repair mechanisms and in maintaining energy homeostasis in photoreceptor cells. The results suggest that the neuroprotective properties of *Sirt1* are gradually attenuated in *rd10* mouse photoreceptor cells (Jaliffa et al. 2009).

Ceramide is a proapoptotic sphingolipid. Retinal ceramide levels are increased in *rd10* mice during the period of maximum photoreceptor death. Single intraocular injections of myriocin, a powerful inhibitor of the biosynthesis of ceramide, lowered retinal ceramide levels to normal values and rescued photoreceptors from apoptotic death. Noninvasive treatment was achieved using eye drops consisting of a suspension of solid lipid nanoparticles loaded with myriocin. Short-term noninvasive treatment lowered retinal ceramide in a manner similar to intraocular injections, indicating that nanoparticles functioned as a vector permitting transcorneal drug administration. Prolonged treatment (10–20 days) with solid lipid nanoparticles increased photoreceptor survival, preserved photoreceptor morphology, and extended the ability of the retina to respond to light as assessed by electroretinography (Strettoi et al. 2010). The treatment consisted of eye drop administration of myriocin in *rd10* mice and showed that the cone photoreceptors, the inner retina, and overall visual performance were preserved well after rod death (Piano et al. 2013).

Meclofenamic acid (MFA) is a potent gap junction blocker that has been used to treat the *rd10* mouse. Investigators hypothesized that blocking gap junctions might eliminate their aberrant activity and restore the light responses in the *rd10* mouse retina. The treatment was shown to increase the sensitivity of the degenerated retina to light stimuli driven by residual photoreceptors. Additionally, it enhanced signal transmission from inner retinal neurons to ganglion cells, potentially allowing the retinal network to preserve the fidelity of signals either from prosthetic electronic devices or from cells optogenetically modified to transduce light (Toychiev et al. 2013).

Activation of the canonical Wnt pathway using exogenous application of Wnt3a leads to functional rescue of photoreceptors in the inherited RD *rd10* mouse. The *rd10* mice were subretinally injected with recombinant Wnt3a protein prior to the onset of RD, and the phenotype was assessed at 7 and 14 days after injection. Wnt3a injection significantly increased levels of the cone photoreceptor marker protein cone transducin and the rod photoreceptor protein rhodopsin. Furthermore, the *rd10* mice injected with β-cateninS33A increased photoreceptor survival by activating Wnt signaling specifically in Müller glia. Retinas injected with β-cateninS33A led to a 44% increase in the number of photoreceptor rows compared with control injections, demonstrating increased photoreceptor survival from Wnt activation. Furthermore, rod and cone photoreceptor ERG responses were also increased, which demonstrates functional rescue of the retina (Patel et al. 2014).

Valosin-containing protein (VCP) is a ubiquitously expressed multifunctional protein, which is a member of the AAA+ (ATPase associated with various activities) protein family. It is a major player causing neurodegeneration. The novel ATPase inhibitors for VCP, with a naphthalene-derived structure, named KUSs (Kyoto University substances) were tested in the *rd10* mouse. KUSs significantly retarded the progres-sion of photoreceptor cell death in *rd10* mutants, and protected the photoreceptor cells morphologically as well as functionally. Reduction of ER stress could be a likely mechanism for KUS-mediated protection of photoreceptors in *rd10* mice (Ikeda et al. 2014).

Adenylyl cyclase (ADCY) plays a role in rod photoreceptor cell death via the phototransduction cascade. The ectopic phototransduction activated by light exposure, which leads to rod photoreceptor cell death, is through the action of transduction (Nakao et al. 2012). SQ 22536, a cell-permeable ADCY inhibitor, was used to treat the *rd10* mouse. A 10 µM/2 µl solution of SQ22536 was injected into the vitreous of one eye and 2 µl PBS into the other eye as a control. The thickness of the retinal ONL was significantly increased by 29% in the retina of the SQ 22536-treated eyes compared with the control eyes (Nakao et al. 2012). The protective effect of light restriction has also been observed and confirmed in the *rd10* mouse (Cronin et al. 2012).

7.3.4 Gene or Cell Replacement Therapies in the *rd10* Mouse Model to Rrescue or Delay Pathological Changes

A number of gene replacement studies have been carried out in the *rd10* model utilizing different delivery systems, all of which demonstrated positive treatment effects on visual outcome. The first study delivered the mouse *Pde6b* gene to the *rd10* mouse retina by an AAV serotype 5 vector (AAV5) containing minimal chicken β-actin promoter/CMV enhancer (smCBA). The purpose of the gene replacement strategy was to determine whether RD could be delayed in the *rd10* model for human autosomal recessive PDEβ-based RP. One eye of a cohort of *rd10* mice raised in a dark environment was subretinally injected at P14 with 1 µl AAV5-smCBA-*Pde6b*. The contralateral eye was not injected. Three weeks post-injection treated *rd10* mice were examined by scotopic and photopic electroretinography and then killed for biochemical and morphologic examination. Substantial scotopic ERG signals were maintained in treated *rd10* eyes, whereas untreated eyes in the same animals showed a minimal response. Treated eyes showed photopic ERG b-wave amplitudes similar to those of the normal eyes, whereas in untreated partner eyes, only 50 % of normal amplitudes remained. Strong PDEβ expression was observed in photoreceptor outer segments of treated eyes. Light microscopy showed a substantial preservation of the ONL in most parts of the treated retina only. Electron microscopy showed good outer segment preservation only in treated eyes (Pang et al. 2008). However, by 6 weeks post treatment, therapeutic effects were no longer evident. In another study, the mouse *Pde6b* gene was delivered subretinally to P14 *rd10* mice by a fast-acting tyrosine-capsid mutant AAV8 (Y733F) vector. sd-OCT, ERG, optomotor behavior tests, and immunohistochemistry showed that AAV8 (Y733F)-mediated PDEβ expression restored retinal function and visual behavior and preserved retinal structure in treated *rd10* eyes for at least 6 months post treatment (Pang et al. 2011). The AAV2/8 and -2/5 vectors alone or in combination with physical (light restriction) and pharmacological agents (calcium channel blocker—nilvadipine) was used in a third study. Subretinal vector administration was performed at P2, 0.75 µl of AAV2/5 or -2/8 encoding βPDE in the right eye. The same dose of AAV2/5- or AAV2/8-CMV-EGFP in 0.75 µl was injected as a control in the left eye. Nilvadipine was administered to *rd10* mice by intraperitoneal injections from P7. The injections were performed once a day (0.05 mg/kg). Dark rearing was performed by moving the late-term pregnant *rd10* females into a continuously dark room until the newborn pups were 24, 28, or 35 days old. Nilvadipine treatment and dark rearing alone were confirmed to delay the RD in the *rd10* mouse. Harvesting of the eyes for histologic analysis was performed at P35. Histologic analysis showed that delivery of AAV2/8 vectors encoding for βPDE resulted in better morphologic rescue than did the AAV2/5 vectors. Also a significant photoreceptor protection after subretinal administration of AAV vectors

encoding EGFP was observed, thus suggesting that injury associated with retinal injection triggers a neurotrophic response in the *rd10* retina (Allocca et al. 2011).

Investigators have also attempted the use of genes other than or in addition to *Pde6b* to rescue the disease phenotype. For example, Deng and coworkers (Deng et al. 2013) delivered the mouse *Pde6c* (phosphodiesterase 6C, cGMP-specific, cone, alpha prime) gene to the *rd10* mouse retina by a fast-acting tyrosine-capsid mutant AAV8 (Y733F) vector. One microliter of AAV8 (Y733F)-smCBA-*Pde6c* was subretinally delivered to P14 *rd10* mice in one eye. The other eye remained uninjected as the control. The subretinal delivery of *Pde6c* rescues rod ERG responses and preserves retinal structure, indicating that cone PDE6C can couple effectively to the rod phototransduction pathway. The restoration of light sensitivity in *rd10* rods is attributable to assembly of cone PDE6C with rod PDE6G (phosphodiesterase 6G, cGMP-specific, rod, and gamma) (Deng et al. 2013). Yao and coworkers (Yao et al. 2012) delivered the X-linked inhibitor of apoptosis (XIAP) gene as an adjunct to mouse *Pde6b* gene to the *rd10* mouse retina by AAV vectors. Injection of AAV5-XIAP alone at P4 and P21 resulted in significant slowing of RD, as measured by outer nuclear thickness and cell counts, but did not result in improved outer segment structure and rhodopsin localization. In contrast, coinjection of AAV5-XIAP and AAV5-PDEβ resulted in increased levels of rescue and decreased rates of RD compared to treatment with AAV5-PDEβ alone. Mice treated with AAV5-XIAP at P4, but not at P21, remained responsive to subsequent rescue by AAV8-733-PDEβ when injected two weeks after moving to a light-cycling environment. Adjunctive treatment with the antiapoptotic gene XIAP confers additive protective effect to gene-replacement therapy with AAV5-PDEβ in the *rd10* mouse. In addition, AAV5-XIAP, when given early, can increase the age at which gene-replacement therapy remains effective, thus effectively prolonging the window of opportunity for therapeutic intervention (Yao et al. 2012).

In retinal diseases, where photoreceptors are eventually lost in entirety, gene therapy is not a viable option and cell replacement may be necessary. The RD in *rd10* and *rd1* mutants was partially rescued by intravitreal injection of a fraction of mouse or human adult bone marrow-derived stem cells (lineage-negative hematopoietic stem cells [Lin-HSCs]) containing endothelial precursors. But the rescued cells after treatment with Lin-HSCs are nearly all cones. Microarray analysis of rescued retinas demonstrates significant upregulation of many anti-apoptotic genes, including small heat shock proteins and transcription factors (Otani et al. 2004, Smith 2004). Other studies using the genetically modified neural stem (NS) cells significantly attenuated photoreceptor degeneration in *rd10* and *rd1* mutants. The neuroprotective effect was significantly more pronounced when clonally derived NS cell lines selected for high expression levels of CNTF were grafted into the *rd1* mouse. Intravitreal transplantations of modified NS cells may thus represent a useful method for preclinical studies aimed at evaluating the therapeutic potential of a cell-based intraocular delivery of NFs in mouse models of photoreceptor degeneration (Jung et al. 2013).

7.4 Summary

The *Pde6b*^{rd10} mouse model of human RP has been widely used in many research laboratories worldwide for studies of human RP and it has been the most commonly used model in the last 10 years. The study of retinal physiology and pathophysiology in the *rd10* model has led to a good understanding of retinal development, maintenance, and function on a molecular, cellular, and tissue-specific level. This knowledge has been used to identify the causative mechanisms in RD and to establish various therapeutic approaches to improve the photoreceptor survival as well as to cure the RD by gene replacement. The results from all the research in the *rd10* mouse have laid the groundwork for the development of an effective therapy to treat patients suffering from RD, hopefully in the near future.

7.4.1 Compliance with Ethical Requirements

Bo Chang and Paul A. Sieving declare that they have no conflict of interest.

Informed Consent and Animal Studies disclosures are not applicable to this review.

Acknowledgments This work has been supported by the National Eye Institute Grant EY019943 and EY011996. I am grateful to Dr. Patsy Nishina and Da Chang for their critical reading and editing of the manuscript.

References

Allocca M, Manfredi A, Iodice C et al (2011) AAV-mediated gene replacement, either alone or in combination with physical and pharmacological agents, results in partial and transient protection from photoreceptor degeneration associated with betaPDE deficiency. Invest Ophthalmol Vis Sci 52(8):5713–5719

Arroba AI, Alvarez-Lindo N, van Rooijen N et al (2011) Microglia-mediated IGF-I neuroprotection in the rd10 mouse model of retinitis pigmentosa. Invest Ophthalmol Vis Sci 52(12):9124–9130

Arroba AI, Alvarez-Lindo N, van Rooijen N et al (2014) Microglia-Müller glia crosstalk in the rd10 mouse model of retinitis pigmentosa. Adv Exp Med Biol 801:373–379

Banerjee P, Kleyn PW, Knowles JA et al (1998) TULP1 mutation in two extended Dominican kindreds with autosomal recessive retinitis pigmentosa. Nat Genet 18:177–179

Barhoum R, Martínez-Navarrete G, Corrochano S et al (2008) Functional and structural modifications during retinal degeneration in the rd10 mouse. Neuroscience 155(3):698–713

Barone I, Novelli E, Piano I et al (2012) Environmental enrichment extends photoreceptor survival and visual function in a mouse model of retinitis pigmentosa. PLoS ONE 7(11):e50726

Barone I, Novelli E, Strettoi E (2014) Long-term preservation of cone photoreceptors and visual acuity in rd10 mutant mice exposed to continuous environmental enrichment. Mol Vis 20:1545–1556

Benkner B, Mutter M, Ecke G et al (2013) Characterizing visual performance in mice: an objective and automated system based on the optokinetic reflex. Behav Neurosci 127(5):788–796

Biswas S, Haselier C, Mataruga A et al (2014) Pharmacological analysis of intrinsic neuronal oscillations in rd10 retina. PLoS ONE 9(6):e99075

Boatright JH, Moring AG, McElroy C et al (2006) Tool from ancient pharmacopoeia prevents vision loss. Mol Vis 12:1706–1714

Bowes C, Li T, Danciger M et al (1990) Retinal degeneration in the rd mouse is caused by a defect in the beta subunit of rod cGMP-phosphodiesterase. Nature 347:677–680

Chang GQ, Hao Y, Wong F (1993) Apoptosis: final common pathway of photoreceptor death in rd, rds, and rhodopsin mutant mice. Neuron 11:595–605

Chang B, Hawes NL, Hurd RE et al (2002) Retinal degeneration mutants in the mouse. Vis Res 42:517–525

Chang B, Hawes N, Davisson M et al (2007a) Mouse models of RP. In: Tombran-Tink J, Barnstable C. (ed) Retinal degenerations: biology, diagnostics, and therapeutics. The Human Press, Inc., New York, pp 149–164

Chang B, Hawes NL, Pardue MT et al (2007b) Two mouse retinal degenerations caused by missense mutations in the beta-subunit of rod cGMP phosphodiesterase gene. Vis Res 47(5):624–633

Chrenek MA, Dalal N, Gardner C et al (2012) Analysis of the RPE sheet in the rd10 retinal degeneration model. Adv Exp Med Biol 723:641–647

Corrochano S, Barhoum R, Boya P et al (2008) Attenuation of vision loss and delay in apoptosis of photoreceptors induced by proinsulin in a mouse model of retinitis pigmentosa. Invest Ophthalmol Vis Sci 49(9):4188–4194

Cronin T, Lyubarsky A, Bennett J (2012) Dark-rearing the rd10 mouse: implications for therapy. Adv Exp Med Biol 723:129–136

Daiger SP, Sullivan LS, Bowne SJ (2013) Genes and mutations causing retinitis pigmentosa. Clin Genet 84:132–141

Deleon E, Lederman M, Berenstein E et al (2009) Alteration in iron metabolism during retinal degeneration in rd10 mouse. Invest Ophthalmol Vis Sci 50(3):1360–1365

Deng WT, Sakurai K, Kolandaivelu S et al (2013) Cone phosphodiesterase-6α′ restores rod function and confers distinct physiological properties in the rod phosphodiesterase-6β-deficient rd10 mouse. J Neurosci 33(29):11745–11753

Doonan F, O'Driscoll C, Kenna P et al (2011) Enhancing survival of photoreceptor cells in vivo using the synthetic progestin norgestrel. J Neurochem 118(5):915–927

Drack AV, Dumitrescu AV, Bhattarai S et al (2012) TUDCA slows retinal degeneration in two different mouse models of retinitis pigmentosa and prevents obesity in Bardet-Biedl syndrome type 1 mice. Invest Ophthalmol Vis Sci 53(1):100–106

Ferrari S, Di Iorio E, Barbaro V et al (2011) Retinitis pigmentosa: genes and disease mechanisms. Curr Genomics 12(4):238–249

Finnegan S, Mackey AM, Cotter TG (2010) A stress survival response in retinal cells mediated through inhibition of the serine/threonine phosphatase PP2A. Eur J Neurosci 32(3):322–334

Galbinur T, Obolensky A, Berenshtein E et al (2009) Effect of para-aminobenzoic acid on the course of retinal degeneration in the rd10 mouse. J Ocul Pharmacol Ther 25(6):475–482

Gargini C, Terzibasi E, Mazzoni F et al (2007) Retinal organization in the retinal degeneration 10 (rd10) mutant mouse: a morphological and ERG study. J Comp Neurol 500(2):222–238

Gibson F, Walsh J, Mburu P et al (1995) A type VII myosin encoded by the mouse deafness gene shaker-1. Nature 374:62–64

Goo YS, Ahn KN, Song YJ et al (2011a) Spontaneous oscillatory rhythm in retinal activities of two retinal degeneration (rd1 and rd10) mice. Korean J Physiol Pharmacol 15(6):415–422

Goo YS, Ahn KN, Song YJ et al (2011b) Comparison of basal oscillatory rhythm of retinal activities in rd1 and rd10 mice. Conf Proc IEEE Eng Med Biol Soc 2011:1093–1096

Guo C, Otani A, Oishi A et al (2012) Knockout of ccr2 alleviates photoreceptor cell death in a model of retinitis pigmentosa. Exp Eye Res 104:39–47

Hagstrom SA, North MA, Nishina PL et al (1998) Recessive mutations in the gene encoding the tubby-like protein TULP1 in patients with retinitis pigmentosa. Nat Genet 18:174–176

Han J, Dinculescu A, Dai X et al (2013) Review: the history and role of naturally occurring mouse models with Pde6b mutations. Mol Vis 19:2579–2589

Heynen SR, Tanimoto N, Joly S et al (2011) Retinal degeneration modulates intracellular localization of CDC42 in photoreceptors. Mol Vis 17:2934–2946

Ikeda HO, Sasaoka N, Koike M et al (2014) Novel VCP modulators mitigate major pathologies of rd10, a mouse model of retinitis pigmentosa. Sci Rep 4:5970

Jae SA, Ahn KN, Kim JY et al (2013) Electrophysiological and histologic evaluation of the time course of retinal degeneration in the rd10 mouse model of retinitis pigmentosa. Korean J Physiol Pharmacol 17(3):229–235

Jaliffa C, Ameqrane I, Dansault A et al (2009) Sirt1 involvement in rd10 mouse retinal degeneration. Invest Ophthalmol Vis Sci 50(8):3562–3572

Jeanny JC, Picard E, Sergeant C et al (2013) Iron and regulatory proteins in the normal and pathological retina. Bull Acad Natl Med 197(3):661–674

Jiang Y, Qi X, Chrenek MA et al (2013) Functional principal component analysis reveals discriminating categories of retinal pigment epithelial morphology in mice. Invest Ophthalmol Vis Sci 54(12):7274–7283

Jiang Y, Qi X, Chrenek MA et al (2014) Analysis of mouse RPE sheet morphology gives discriminatory categories. Adv Exp Med Biol 801:601–607

Jung G, Sun J, Petrowitz B et al (2013) Genetically modified neural stem cells for a local and sustained delivery of neuroprotective factors to the dystrophic mouse retina. Stem Cells Transl Med 2(12):1001–1010

Karlstetter M, Ebert S, Langmann T (2010) Microglia in the healthy and degenerating retina: insights from novel mouse models. Immunobiology 215(9/10):685–691

Keeler C (1924) The inheritance of a retinal abnormality in white mice. Proc Natl Acad Sci U S A 10:329

Keeler C (1966) Retinal degeneration in the mouse is rodless retina. J Hered 57(2):47–50

Keen TJ, Inglehearn CF (1996) Mutations & polymorphisms in the human peripherin-RDS gene & their involvement in inherited retinal degeneration. Hum Mut 8:297–303

Komeima K, Rogers BS, Campochiaro PA (2007) Antioxidants slow photoreceptor cell death in mouse models of retinitis pigmentosa. J Cell Physiol 213(3):809–815

Lee SY, Usui S, Zafar AB et al (2011) N-Acetylcysteine promotes long-term survival of cones in a model of retinitis pigmentosa. J Cell Physiol 226(7):1843–1849

Mazzoni F, Novelli E, Strettoi E (2008) Retinal ganglion cells survive and maintain normal dendritic morphology in a mouse model of inherited photoreceptor degeneration. J Neurosci 28(52):14282–14292

McLaughlin ME, Sandberg MA, Berson EL et al (1993) Recessive mutations in the gene encoding the beta-subunit of rod phosphodiesterase in patients with retinitis pigmentosa. Nat Genet 4:130–134

McLaughlin ME, Ehrhart TL, Berson EL et al (1995) Mutation spectrum of the gene encoding the beta subunit of rod phosphodiesterase among patients with autosomal recessive retinitis pigmentosa. Proc Natl Acad Sci U S A 92:3249–3253

Menzler J, Channappa L, Zeck G (2014) Rhythmic ganglion cell activity in bleached and blind adult mouse retinas. PLoS ONE 9(8):e106047

Mitton KP, Guzman AE, Deshpande M et al (2014) Different effects of valproic acid on photoreceptor loss in Rd1 and Rd10 retinal degeneration mice. Mol Vis 20:1527–1544

Muir ER, De La Garza B, Duong TQ (2013) Blood flow and anatomical MRI in a mouse model of retinitis pigmentosa. Magn Reson Med 69(1):221–228

Murakami Y, Ikeda Y, Yoshida N et al (2012a) MutT homolog-1 attenuates oxidative DNA damage and delays photoreceptor cell death in inherited retinal degeneration. Am J Pathol 181(4):1378–1386

Murakami Y, Matsumoto H, Roh M et al (2012b) Receptor interacting protein kinase mediates necrotic cone but not rod cell death in a mouse model of inherited degeneration. Proc Natl Acad Sci U S A 109(36):14598–14603

Nakao T, Tsujikawa M, Notomi S et al (2012) The role of mislocalized phototransduction in photoreceptor cell death of retinitis pigmentosa. PLoS ONE 7(4):e32472

Noben-Trauth K, Naggert JK, North MA et al (1996) A candidate gene for the mouse mutation tubby. Nature 380:534–538

Obolensky A, Berenshtein E, Lederman M et al (2011) Zinc-desferrioxamine attenuates retinal degeneration in the rd10 mouse model of retinitis pigmentosa. Free Radic Biol Med 51(8):1482–1491

O'Driscoll C, Doonan F, Sanvicens N et al (2011) A novel free radical scavenger rescues retinal cells in vivo. Exp Eye Res 93(1):65–74

Ohnaka M, Miki K, Gong YY et al (2012) Long-term expression of glial cell line-derived neurotrophic factor slows, but does not stop retinal degeneration in a model of retinitis pigmentosa. J Neurochem 122(5):1047–1053

Otani A, Dorrell MI, Kinder K et al (2004) Rescue of retinal degeneration by intravitreally injected adult bone marrow-derived lineage-negative hematopoietic stem cells. J Clin Invest 114(6):765–774

Oveson BC, Iwase T, Hackett SF et al (2011) Constituents of bile, bilirubin and TUDCA, protect against oxidative stress-induced retinal degeneration. J Neurochem 116(1):144–153

Pang JJ, Boye SL, Kumar A et al (2008) AAV-mediated gene therapy for retinal degeneration in the rd10 mouse containing a recessive PDEbeta mutation. Invest Ophthalmol Vis Sci 49(10):4278–4283

Pang JJ, Dai X, Boye SE et al (2011) Long-term retinal function and structure rescue using capsid mutant AAV8 vector in the rd10mouse, a model of recessive retinitis pigmentosa. Mol Ther 19(2):234–242

Park H, Tan CC, Faulkner A et al (2013) Retinal degeneration increases susceptibility to myopia in mice. Mol Vis 19:2068–2079

Patel AK, Surapaneni K, Yi H et al (2014) Activation of Wnt/β-catenin signaling in Muller glia protects photoreceptors in a mouse model of inherited retinal degeneration. Neuropharmacology pii: S0028-3908(14):00431–00436

Peng B, Xiao J, Wang K et al (2014) Suppression of microglial activation is neuroprotective in a mouse model of human retinitis pigmentosa. J Neurosci 34(24):8139–8150

Pennesi ME, Michaels KV, Magee SS et al (2012) Long-term characterization of retinal degeneration in rd1 and rd10 mice using spectral domain optical coherence tomography. Invest Ophthalmol Vis Sci 53(8):4644–4656

Phillips MJ, Walker TA, Choi HY et al (2008) Tauroursodeoxycholic acid preservation of photoreceptor structure and function in the rd10 mouse through postnatal day 30. Invest Ophthalmol Vis Sci 49(5):2148–2155

Phillips MJ, Otteson DC, Sherry DM (2010) Progression of neuronal and synaptic remodeling in the rd10 mouse model of retinitis pigmentosa. J Comp Neurol 518(11):2071–2089

Phillips MJ, Otteson DC, Sherry DM (2011) Progression of neuronal and synaptic remodeling in the rd10 mouse model of retinitis pigmentosa. J Cell Mol Med 15(8):1778–1787

Piano I, Novelli E, Gasco P et al (2013) Cone survival and preservation of visual acuity in an animal model of retinal degeneration. Eur J Neurosci 37(11):1853–1862

Picard E, Jonet L, Sergeant C et al (2010) Overexpressed or intraperitoneally injected human transferrin prevents photoreceptor degeneration in rd10 mice. Mol Vis 16:2612–2625

Pittler SJ, Keeler CE, Sidman RL et al (1993) PCR analysis of DNA from 70-year-old sections of rodless retina demonstrates identity with the mouse rd defect. Proc Natl Acad Sci U S A 90(20):9616–9619

Portera-Cailliau C, Sung CH, Nathans J et al (1994) Apoptotic photoreceptor cell death in mouse models of retinitis pigmentosa. Proc Natl Acad Sci U S A 91(3):974–978

Rex TS, Allocca M, Domenici L et al (2004) Systemic but not intraocular Epo gene transfer protects the retina from light-and genetic-induced degeneration. Mol Ther 10(5):855–861

Rodríguez-Muela N, Hernández-Pinto AM, Serrano-Puebla A et al (2014) Lysosomal membrane permeabilization and autophagy blockade contribute to photoreceptor cell death in a mouse model of retinitis pigmentosa. Cell Death Differ. doi:10.1038/cdd.2014.203

Rösch S, Johnen S, Müller F et al (2014) Correlations between ERG, OCT, and Anatomical Findings in the rd10 Mouse. J Ophthalmol 2014:874751

Samardzija M, Neuhauss S, Joly S et al (2010) Animal models for retinal degeneration. In Pang I, Clark AF (eds) Animal models for retinal diseases. The Human Press, Inc., New York, pp 51–79

Samardzija M, Wariwoda H, Imsand C et al (2012) Activation of survival pathways in the degenerating retina of rd10 mice. Exp Eye Res 99:17–26

Smith LE (2004) Bone marrow-derived stem cells preserve cone vision in retinitis pigmentosa. J Clin Invest 114(6):755–757

Stasheff SF, Shankar M, Andrews MP (2011) Developmental time course distinguishes changes in spontaneous and light-evoked retinal ganglion cell activity in rd1 and rd10 mice. J Neurophysiol 105:3002–3009

Strettoi E, Gargini C, Novelli E et al (2010) Inhibition of ceramide biosynthesis preserves photoreceptor structure and function in a mouse model of retinitis pigmentosa. Proc Natl Acad Sci U S A 107(43):18706–18711

Thomas BB, Shi D, Khine K et al (2010) Modulatory influence of stimulus parameters on optokinetic head-tracking response. Eur J Neurosci Neurosci Lett 479(2):92–96

Toychiev AH, Ivanova E, Yee CW et al (2013) Block of gap junctions eliminates aberrant activity and restores light responses during retinal degeneration. J Neurosci 33(35):13972–13977

Travis GH, Brennan MB, Danielson PE et al (1989) Identification of a photoreceptor-specific mRNA encoded by the gene responsible for retinal degeneration slow (rds). Nature 338:70–73

Uren PJ, Lee JT, Doroudchi MM et al (2014) A profile of transcriptomic changes in the rd10 mouse model of retinitis pigmentosa. Mol Vis 20:1612–1628

Usui S, Komeima K, Lee SY et al (2009) Increased expression of catalase and superoxide dismutase 2 reduces cone cell death in retinitis pigmentosa. Mol Ther 17(5):778–786

Usui S, Oveson BC, Iwase T et al (2011) Overexpression of SOD in retina: need for increase in H2O2-detoxifying enzyme in same cellular compartment. Free Radic Biol Med 51(7):1347–1354

Wang K, Xiao J, Peng B et al (2014) Retinal structure and function preservation by polysaccharides of wolfberry in a mouse model of retinal degeneration. Sci Rep 4:7601

Weil D, Blanchard S, Kaplan J et al (1995) Defective myosin VIIA gene responsible for Usher syndrome type 1B. Nature 374:60–61

Wisard J, Chrenek MA, Wright C et al (2010) Non-contact measurement of linear external dimensions of the mouse eye. J Neurosci Methods 187(2):156–166

Won J, Shi LY, Hicks W et al (2011) Mouse model resources for vision research. J Ophthalmol 2011:391384

Yao J, Jia L, Khan N et al (2012) Caspase inhibition with XIAP as an adjunct to AAV vector gene-replacement therapy: improving efficacy and prolonging the treatment window. PLoS ONE 7(5):e37197

Yoshida N, Ikeda Y, Notomi S et al (2013) Laboratory evidence of sustained chronic inflammatory reaction in retinitis pigmentosa. Ophthalmology 120(1):e5–e12

Zieger M, Ahnelt PK, Uhrin P (2014) CX3CL1 (fractalkine) protein expression in normal and degenerating mouse retina: in vivo studies. PLoS ONE 9(9):e106562

Commentary

Paul A. Sieving
email: pas@NEI.NIH.GOV
National Eye Institute,
National Institutes of Health,
Bethesda, MD, USA

RP is the clinical term applied to conditions which cause progressive dysfunction and ultimately death of photoreceptor cells. Rod photoreceptors are particularly vulnerable to loss and ultimately die first, but cone photoreceptors eventually fail also. With loss of both rods and cones, affected individuals are rendered essentially blind.

The breakthrough in understanding the biological causes of RP came in 1990 with identification of mutations in the rhodopsin gene as causing progressive blindness in a large Irish family in which autosomal dominant inheritance caused successive generations to be affected. Other causative genes were subsequently identified for other families, and currently nearly 100 different genes are associated with human RP. On this basis, we now know that the majority of these progressively blinding conditions are caused by mutations in single genes, and the condition is properly understood to be a constellation of monogenic photoreceptor neurodegenerative diseases.

Many of these RP conditions have now been mimicked in mouse models through transgenic laboratory techniques to create mutations in the mouse gene corresponding to the human condition. These models present a remarkable opportunity to study the inciting causes and to investigate the pathophysiology causing progressive rod and cone failure leading to cell death. In turn, investigators are now pursuing opportunities to intervene therapeutically and rescue the vision, first of affected mice and subsequently, we hope, of human individuals with the corresponding conditions.

The author of this chapter, Bo Chang, presents an in-depth analysis of one model among the nearly 100 different genetic models that have been established for RDs. Bo Chang has deep expertise in the rd10 mouse that carries a mutation in the *Pde6b* gene. While at first it might seem that a chapter focused on a single model has rather limited scope. However, this approach is eminently sensible. From the encyclopedic inventory of therapeutic strategies that have been employed to rescue photoreceptors of the rd10 mouse, the readers will be aware of the considerable work involved in understanding cellular pathophysiology that underpins any therapeutic attempts to alleviate vision loss.

The rd10 model is said to exhibit "slow progression," but in fact the retina loses all rod and cone photoreceptors by age 6–9 weeks, corresponding to human age of 20–30 years old. In general, humans with RP still have considerable vision at this age. Hence the term "slow" must be understood in relation to rd1 (*Pde6a* mutation) mice which lose all rods before 28 days age and cones soon thereafter. This illustrates the genetic

and pathophysiologic diversity of RP diseases across the many animal models and the corresponding human conditions. An advantage of more rapid mouse models of RP, such as rd10, is that they can be studied at the "calendar-speed" of an operating laboratory. Even so, a proper laboratory therapy study will run for 1 or 2 years for a single compound.

As the chapter points out, the same tools used to study human vision clinically can also be employed with mouse models in vivo. These include direct viewing of the retina either through a clinical ophthalmoscope or with the camera lens. More detailed structural analysis is provided using optical coherence tomography to view the layers of the intact retina in vivo, in mouse or human. On a functional scale, visual acuity can be measured in mice with a moving bar pattern as is also used to assess infant human vision. In addition, the ERG provides a detailed cellular-level analysis of photoreceptor activity and neural transmission into the proximal retina, in mouse models and in human patients. This means that information developed from therapeutic studies in mouse can be translated to human outcomes for intervention trials. This greatly simplifies the regulatory interface with the FDA for gaining approval for human studies with RP.

This chapter provides an extensive review of the many ways that biological knowledge has been developed regarding the causes and cellular consequences of the abnormal Pde6b protein in the rd10 mouse. And of equal interest, this chapter reviews the diverse biological strategies that have been employed to rescue vision by slowing or ameliorating the pathophysiology. At this point in history much is known about RP disease in mice. Despite this, attempts at human therapy for RP remain extremely limited to date. We can hope this will soon change. And when it does, it will be attributed in great degree to the availability of the many, precise genetic mouse models that mimic these human monogenic photoreceptor neurodegenerative conditions with great fidelity.

An Animal Model of Graves' Orbitopathy

8

J. Paul Banga, Sajad Moshkelgosha, Utta Berchner-Pfannschmidt and Anja Eckstein

8.1 Introduction

Thyroid autoimmune diseases represent one of the most common autoimmune conditions affecting humans in the western world, with a prevalence rate of 1–2% in the female Caucasian population (Jacobson et al. 1997). Among the autoimmune conditions, Graves' disease arises as a consequence of the induction of thyroid stimulating antibodies with strong agonist activity to the thyroid stimulating hormone (TSH) receptor (TSHR) resulting in hyperthyroidism (Weetman 2000). The receptor is expressed predominantly in the thyroid follicular cells where stimulation by the pituitary hormone TSH leads to the production of thyroid hormone. The TSHR is of large interest to immunologists, as it is one of the few receptors to which immunological tolerance is commonly disrupted leading to autoimmune disease (Rapoport et al. 1998). The inductive events leading to the production of the thyroid stimulating antibodies in Graves' disease remain completely unknown, although both genetic background and environmental factors are known to play contributory roles in the etiology of the disease (Tomer and Davies 2007; Vanderpump 2011). Extrathyroidal expression of functional TSHR has also been found in other tissues, principally in orbital fibroblasts and adipocytes (Davies 1994; Bahn 2010), while the expression in bone tissue remains controversial (Abe et al. 2003; Baliram et al. 2012; Duncan Bassett et al. 2008).

Common manifestations accompanying Graves' disease include extrathyroidal conditions of inflammatory eye disease known as Graves' orbitopathy (GO) (also referred to as thyroid eye disease or exophthalmos) and dermopathy (Weetman 2000). Graves' orbitopathy occurs in almost all patients with Graves' disease when examined by magnetic resonance imaging (MRI, Villadolid et al. 1995), clinically overt disease occurs in 40–50% patients, with around 5% patients where it may be sight threatening (Kendall-Taylor and Perros 1998). The condition is characterized by expansion of the orbital fat tissue resulting principally from adipogenesis, deposition of high-molecular weight glycosaminoglycans such as hyaluronan and fibrosis (Bahn 2010). The events that link orbital inflammation in GO with the ongoing autoimmune process

J. P. Banga (✉) · S. Moshkelgosha
The Rayne Institute, King's College London,
123 Coldharbour Lane, London SE5 9NU, UK
e-mail: paul.banga@kcl.ac.uk

J. P. Banga · S. Moshkelgosha · U. Berchner-
Pfannschmidt · A. Eckstein
Universitäts-Augenklinik, Gruppe für Molekulare
Ophthalmologie, Medizinisches Forschungzentrum,
Universität Duisburg-Essen, Hufelandstrasse 55,
45122 Essen, Germany

A. Eckstein
e-mail: anja.eckstein@uk-essen.de

© Springer International Publishing Switzerland 2016
C.-C. Chan (ed.), *Animal Models of Ophthalmic Diseases*, Essentials in Ophthalmology,

Table 8.1 Evidence for Graves' orbitopathy as an autoimmune disease

Inflammatory cell infiltrates in the orbital tissue of CD3[+] T cells and macrophages
Pathogenic autoantibodies to the thyrotropin receptor
Chemosis characterized by histopathological signs of inflammatory, congested blood vessels, and accumulation of mast cells
Association in family members favors contribution of genetic factors
Success of immunotherapies targeting immune cells, e.g., CD20 on B cells by rituximab

occurring in the thyroid gland remain to be clarified (Table 8.1). The fact that both the thyroid follicular cells and the orbital fibroblasts and adipocytes constitutively express TSHR may be the common denominator, with the autoimmune reaction to the thyroidal receptor spreading to the orbital tissue (Bahn 2010; Wall 2014).

To define mechanisms underlying GO requires experimental models that manifest the clinical course and pathology of the disease. However, no spontaneous models of Graves' disease and the associated orbital manifestations similar to GO are available, as these conditions appear to occur spontaneously only in humans (Aliesky et al. 2013). Even nonhuman primates, species that are mostly similar and evolutionary closest to humans, do not appear to succumb to spontaneous autoimmune hyperthyroidism (McLachlan et al. 2011). Since in humans, GO occurs commonly in patients with ongoing autoimmune Graves' hyperthyroid disease, any development of a model for the orbital condition has to have a corollary of a model for experimental Graves' disease.

Mouse models of induced Graves' disease have been developed over the past two decades, and although some reported orbital manifestations, none have been verified in independent laboratories and have proved to be disappointing failures (Baker et al. 2005; Dağdelen et al. 2009; Wiesweg et al. 2013). More recently, the adoptive transfer of TSHR primed spleen cells from TSHR knockout mice (which fail to develop tolerance

to the TSHR) to athymic nu-/nu- mice showed small histopathological lesions in the orbital tissue of some animals, but clinically overt disease was not present (Nakahara et al. 2012).

In this chapter, we discuss the development and latest research on a mouse model of GO that recapitulates a number of features to those present in human disease. However, it becomes necessary to briefly describe mouse models of Graves' disease to comprehend the development of the model for GO. For more detail on models of Graves' disease, the reader is referred to recent reviews that describe the multiple models, which recapitulate different aspects of the disease (Dağdelen et al. 2009; Wiesweg et al. 2013).

8.2 Experimental Models of Graves' Disease

Generally, experimentally induced models of autoimmune disease are induced by immunization with the potential self-antigen implicated in the autoimmune condition together with an adjuvant (termed active induction) in genetically susceptible animals (Rose 1988). Such methods allow studies on both genetic and molecular investigations, as well as T cell regulation, and have added much to the understanding of autoimmune conditions (Kong 1997). However, initial attempts to establish experimental Graves' disease by immunization with characterized preparations of purified human TSHR (hTSHR) protein in an adjuvant did not lend to the induction of thyroid stimulating antibodies and hyperthyroid disease, although large amounts of antibody and T cell reactivity to the receptor were produced (Carayanniotis et al. 1995; Wang et al. 1998; Banga 2007).

In vivo delivery of TSHR was the key to developing successful mouse models of Graves' disease. Several induction models for Graves' disease in inbred mice have been described, with different disease incidence, course, and pathological features (Dağdelen et al. 2009; Wiesweg et al. 2013).

Almost all mouse models of Graves' disease have relied on the BALB/c (H-2d) inbred mouse strain with the highest degree of disease susceptibility (Dağdelen et al. 2009; Wiesweg et al. 2013). To establish efficient in vivo expression of TSHR, genetic delivery by plasmids encoding the extracellular region of hTSHR, termed hTSHR A-subunit, was evaluated. However, these early plasmid DNA delivery models proved to be unreliable in terms of low disease incidence and reproducibility (Pichurin et al. 2001; Rao et al. 2003; Banga 2007). The most commonly used model, replicated in a number of laboratories, relied on adenovirus-mediated delivery of hTSHR A-subunit cDNA (Chen et al. 2003), originally based on the innovative findings of Nagayama and colleagues using a full-length hTSHR coding plasmid for immunization (Nagayama et al. 2002). The earlier mouse models of Graves' disease, including the adenovirus model, did not lead to any visible signs of overt GO (Nagayama et al. 2002; Gilbert et al. 2006). This raised questions whether other autoantigens, such as the insulin growth factor receptor 1 (IGF-1R), by itself or in association with TSHR was the reason for the lack of orbital pathology (Wiersinga 2012). A notable feature of Graves' disease models is short-lived immune responses, as the anti-TSHR response declines soon after the cessation of immunization (Gilbert et al. 2006; McLachlan et al. 2012). It is likely that in these models the short duration of anti-TSHR responses results in acute phase of the disease, which is sufficient to induce hyperthyroid disease, but inadequate for the development of pathological events leading to remodeling of orbital tissue. Thus, development of protocols to induce a long and sustained immune response to TSHR in animals undergoing experimental Graves' disease leading to chronic hyperthyroidism was our rationale for attempts to develop orbital manifestations in the model.

8.3 Experimental Model of Graves' Orbitopathy

Initial Attempts to Induce Chronic Hyperthyroidism in Mice Earlier studies attempted to produce chronic stimulation of the thyroid gland by growing hybridoma cells secreting thyroid stimulating antibody in vivo in the peritoneal cavity of mice, but have the drawback that it is difficult to know the amount of monoclonal antibodies (mAb) available in vivo, nor the degree of TSHR stimulation resulting in hyperthyroidism (Ando et al. 2004). Our studies of atypical chronic hyperthyroid disease relied on regular weekly intravenous transfer over several weeks, of microgram quantities of mAb with strong thyroid stimulating activity (Gilbert at al. 2006; Flynn et al. 2007). These studies in humanized HLA-DRB1*0301 transgenic mice in the NOD genetic background resulted in achieving the casual relationship between chronic stimulation and resultant hyperthyroidism and goiter, but no visible signs of orbital manifestations were apparent (Flynn et al. 2007).

Genetic delivery of hTSHR plasmid by in vivo electroporation reported the longevity of the induced immune response to hTSHR (Kaneda et al. 2007). We refined this protocol using an hTSHR A-subunit plasmid with caliper electrodes for close-field electroporation by delivery of brief, intense electrical pulses to shaved quadricep leg muscle and surrounding tissue which leads to temporary opening of the cell membrane to allow efficient plasmid delivery to inside the cell (Widera et al. 2000). This approach resulted in a sustained response to TSHR, together with high incidence of induced hyperthyroidism in female BALB/c mice (Zhao et al. 2011). All immune animals developed a strong thyroid stimulating antibody response to hTSHR, accompanied with a large proportion (75%) of animals with hyperthyroidism. Interestingly, antibodies induced to hTSHR in other murine models of experimental Graves' disease cross-react weakly with the autologous mouse TSHR, but the special subset of induced thyroid stimulating antibodies cross-react strongly with the stimulating "trigger" determinants on the mouse TSHR to activate the receptor in vivo leading to hyperthyroidism (Nakahara et al. 2010). Histopathological examination of the orbital tissue gave the first indications of orbital inflammation and fibrosis in the model (Zhao et al. 2011). In principle, our model replicates the situation in Graves' disease patients by an ongoing autoimmune response and continuous chronic stimulation of thyroid gland.

IGF-1R has been implicated as a second antigen in GO. Autoantibodies to IGF-1R have been identified in patients with GO (Weightman et al. 1993), although other recent intensive searches for these antibodies have failed to support these findings (Wiersinga 2012; Minich et al. 2013; Varewijck et al. 2013). Our studies using an IGF-1R α-subunit coding plasmid for immunization by close-field electroporation in BALB/c mice failed to induce any pathology in immune animals, although robust anti-IGF-1R antibodies were induced (Zhao et al. 2011; Moshkelgosha et al. 2013). Thus, the role of IGF-1R in GO remains challenging. Interestingly, recent studies provide evidence for receptor cross-talk between G-protein coupled receptors and tyrosine kinase receptors (Gavi et al. 2006). Indeed, such cross-talk has recently been reported between TSHR and IGF-1R in orbital fibroblasts that may provide an explanation for the discrepant findings on anti-IGF-1R antibodies in GO (Krieger et al. 2015).

Development of a Bona-Fide Model of Graves' Orbitopathy In a further development of the electroporation model of experimental Graves' disease (Zhao et al. 2011), we evaluated small changes in the delivery of the plasmid inoculum into the leg muscle and whether these changes correlated with orbital lesions. We also improved the histopathology of the orbital region of the mouse (Smith 2002), by using the optic nerve as an anatomical reference marker in the section (Fig. 8.1, Panel a) (Johnson et al. 2013). The modifications to the model presented with the development of GO in mice undergoing autoimmunity to hTSHR (Moshkelgosha et al. 2013).

The pathology of GO in the model is primarily driven by expansion of the orbital tissue by inflammation and adipogenesis. The majority of the animals show an inflammatory infiltrate into the extraocular muscle (Fig. 8.1, Panel b). In addition, orbital pathology in a few animals was characterized by inflammatory T cells around the optic nerve, but without infiltrate in the optic nerve or in the surrounding retrobulbar tissue (Fig. 8.1, Panels c and d). There were also differences, as some animals exhibit expansion of the retrobulbar adipose tissue, which is sufficiently large enough to separate the orbital muscle fiber bundles (Fig. 8.1, Panels e and f). Immunohistochemical staining identified CD3$^+$ T cells (Fig. 8.1, Panels g and h). There was a distinct absence of B cells in the orbital infiltrate. End-stage disease was characterized by fibrosis (Fig. 8.1, Panel i). Most striking, however, was the finding that some animals also show extra-orbital signs of inflammation, such as chemosis (Fig. 8.1, Panel j) with congested blood vessels and accumulation of mast cells (Fig. 8.1, Panels k and l). Further examination of soft tissue in the orbital region by small animal MRI in mice undergoing experimental GO shows clear hypertrophy of the orbital muscles and proptosis of the eye (not shown) (Moshkelgosha et al. 2013). Thus, the pathology of GO model closely mimics the disease heterogeneity reminiscent of the remodeling of orbital tissue in the human condition (Table 8.2).

The orbital remodeling in animals undergoing experimental GO was associated with sustained antibody responses to TSHR during the entire course of the disease. The antibodies induced to the TSHR comprise of both thyroid stimulating and thyroid stimulating blocking antibodies and depending on the delivery of the hTSHR plasmid inoculum, results predominantly with immune animals presenting with hyperthyroid disease (Zhao et al. 2011) or both hyperthyroid or hypothyroid disease (Moshkelgosha et al. 2013). Importantly, our studies emphasize that experimental GO develops under hyperthyroid or hypothyroid conditions, very similar to the situation in human patients (Table 8.2).

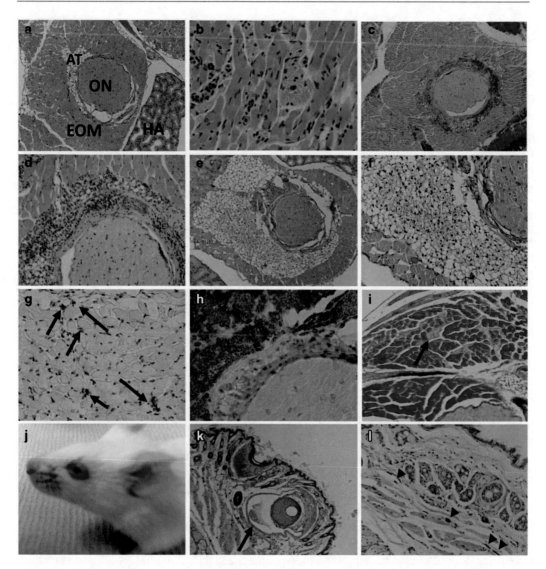

Fig. 8.1 Representative H&E stained sections of orbital tissue from mice undergoing Graves' ophthalmopathy, following immunization with the hTSHR A-subunit plasmid by close-field electroporation. **a** Orbital tissue from control b-Gal plasmid immunized mouse (×100). *EOM* extraocular muscle; *ON* optic nerve; *AT* adipose tissue; *HA* harderian gland. **b** Inflammatory infiltrate into orbital muscle tissue (×100). **c** Perineural infiltrate around optic nerve (×100). **d** Higher magnification (×200). **e** Expansion of orbital fat tissue (×100). **f** Higher magnification (×200). **g** immunohistochemistry (IHC) to show CD3+T cells (*arrowed*) into orbital muscle tissue (×200). **h** IHC showing CD3 + T cells in the perineural infiltrate around optic nerve (×100). **i** Fibrosis in retrobulbar tissue at end stage of disease, Masson's Trichrome-stained section (×100). **j** Mouse head to show animal undergoing chemosis. **k** Chemosis eyelid to show congested blood vessels, *arrowed* (×100). **l** Higher magnification to show accumulation of mast cells, *arrow heads* (×200)

Table 8.2 Clinical outcome in Graves' orbitopathy model—similarities and differences from the human condition

Similarities
Inflammatory cell infiltrates in the orbital tissue of CD3$^+$ CD4$^+$ T cells, macrophages, and mast cells (80% diseased animals)
Adipogenesis, expansion of orbital adipose tissue (10% diseased animals)
Fibrosis of orbital muscle tissue
Presence of thyrotropin receptor antibodies, with activating or blocking antibodies
Hyperthyroid or hypothyroid status in animals undergoing Graves' orbitopathy
Presence of antibodies to IGF-1R
Eyelid inflammation, chemosis
Orbital muscle hypertrophy leading to proptosis
Differences
Inflammatory T cells around the optic nerve (5% diseased animals), but without infiltrate in the optic nerve
Absence of thyroid inflammation
Unilateral disease not found, both eyes show the same pathology

8.4 Concluding Remarks

The development of a model for GO has been dependent on immunization by electroporation with hTSHR plasmid, providing compelling evidence on the TSHR as the primary autoantigen in the disease. The findings also provide an explanation for the link between thyroid autoimmunity and orbital remodeling in GO (Table 8.1), with TSHR as the common antigen present in thyroid follicular cells and orbital fibroblasts. Recent studies indicate that the thyroid stimulating antibody binding may be dependent on the quaternary structure of TSHR A-subunit assembling as higher subunit oligomers in the form of trimers (Chen et al. 2015). This may explain the molecular basis of dependence on either the cell-associated receptor or the genetic delivery of TSHR A-subunit for induction of Graves' disease and associated complications of GO.

The pathological features of orbital lesions are most reminiscent to those in human patients with GO, characterized by an inflammatory infiltrate of CD3$^+$ T cells and macrophages or expansion of adipose tissue. It has been hypothesized that the orbital lesions primarily result from the activity of T cells and macrophages as antigen-presenting cells, leading to cytokine production, followed by the cascade of orbital remodeling (Bahn 2010). Moreover, some animals show characteristic signs of chemosis, accompanied by eyelid pathology of dilated and congested blood vessels. We have discussed the importance of chronic stimulation of TSHR as one of the critical factors in the induction of GO in the model.

Early studies suggest that the GO model by close-field electroporation has been successfully repeated in independent laboratories of the authors; the data will be published separately. Nevertheless, it is important to recognize the limitations of the experimental model, as the model is not simple and has a long time frame for development (Moshkelgosha et al. 2013; Bahn 2013). This is likely to be related to the establishment of chronic hyperthyroidism for manifestation of orbital complications. Genetic predisposition to GO in the model is presently not known, as only one inbred strain (BALB/c, H-2d) with susceptibility has been evaluated. It will be important to evaluate other inbred strains, particularly C57/BL6 (H-2b) mice, for induction of GO, if transgenic and immune response gene knockout and knockin animals are to be evaluated to begin to understand the immunological basis of the condition. The new model of GO will be invaluable to understand the natural history of the disease and events that underlie the continued progression of the autoimmune response to the orbital components in GO. Finally, for treating this difficult condition (Bartalena 2013), rather than evaluating the efficacy of therapeutic immunomodulators in expensive clinical trials, such as the randomized controlled trials reported recently for rituximab (Stan et al. 2014; Salvi et al. 2014) or the ongoing Phase 2 multicenter control trial with humanized anti-IGF-1R blocking monoclonal antibody, teprotumumab (www.clinicaltrials.gov/ct2/show/record/NCT01868997), the new model can be applied relatively easily to study the benefits and mechanistic pathways of therapeutic agents.

8.5   Checklists and Bullet Points of Core Messages

- TSHR as the primary target antigen in GO
- Orbital enlargement in the mouse model defined as predominantly muscle inflammation and hypertrophy or by expansion of adipose tissue.
- Pathological findings involve infiltration of orbital muscle tissue with CD4$^+$ T cells and macrophages as antigen-presenting cells.
- Extrathyroidal complication of chemosis with dilated and congested orbital blood vessels in eyelids.
- Hyperthyroid or hypothyroid status characterized by thyroid stimulating or blocking antibodies, respectively, leads to induction of GO.
- Role of antibodies in IGF-1R remains unanswered.

8.6   Compliance with Ethical Requirements

J. Paul Banga, SajadMoshkelgosha, UttaBerchner-Pfannschmidt, Anja Eckstein, Rebecca Bahn, Shivani Gupta, and Raymond S. Douglas declare that they have no conflict of interest.

No human studies were carried out by the authors for this article.

All institutional and national guidelines for the care and use of laboratory animals were followed.

All the animal studies reported in the manuscript were carried out in King's College London, UK under license from UK Home Office with strict standards of humane animal care.

References

Abe E et al (2003) TSH is a negative regulator of skeletal remodeling. Cell 115:151–162

Aliesky H et al (2013) Thyroid autoantibodies are rare in nonhuman great apes and hypothyroidism cannot be attributed to thyroid autoimmunity. Endocrinology 154:4896–4907

Ando T et al (2004) Concentration dependent regulation of thyrotropin receptor function by thyroid stimulating antibody. J Clin Invest 113:1589–1595

Bahn RS (2010) Graves' ophthalmopathy. N Engl J Med 362:726–738

Bahn RS (2013) News and views: at long last, an animal model of Graves' orbitopathy. Endocrinology 154:2989–2991

Baker G et al (2005) Reevaluating thyrotropin receptor-induced mouse models of Graves' disease and ophthalmopathy. Endocrinology 146:835–844

Balliram R et al (2012) Hyperthyroid-associated osteoporosis is exacerbated by the loss of TSH signaling. J Clin Invest 122:3737–3741

Banga JP (2007) The long and winding road for an experimental model of hyperthyroid Graves' disease. In: Wiersinga WM, Drexhage HA, Weetman AP (eds) The thyroid and autoimmunity. Georg Thieme Verlag, Stuttgart, pp 118–125

Bartalena L (2013) Graves' orbitopathy: imperfect treatments for a rare disease. Eur Thyroid J 2:259–269

Carayanniotis G et al (1995) Unaltered thyroid function in mice responding to a highly immunogenic thyrotropin receptor: implications for the establishment of a mouse model for Graves' disease. Clin Exp Immunol 99:294–302

Chen CR et al (2003) Thethyrotropin receptor autoantigen in Graves' disease is the culprit as well as the victim. J Clin Invest 111:1897–1904

Chen CR et al (2015) Crystal structure of a TSH receptor monoclonal antibody: insight into Graves' disease pathogenesis. Mol Endocrinol 29:99–107

Dağdelen S et al (2009) Toward better models of hyperthyroid Graves' disease. Endocrinol Metab Clin North America 38:343–354

Davies TF (1994) Thethyrotropin receptors spread themselves around. J Clin Endocrinol Metab 79:1232–1233

Duncan Bassett JH et al (2008) A lack of thyroid hormone rather than excess thyrotropin causes abnormal skeletal development in hypothyroidism. Mol Endocrinol 22:501–512

Flynn JC et al (2007) Chronic exposure in vivo to thyrotropin receptor stimulating monoclonal antibodies sustains high thyroxine levels and thyroid hyperplasia in thyroid autoimmunity-prone HLA-DRB1*0301 transgenic mice. Immunology 122:261–267

Gavi S et al (2006) G-protein coupled receptors and tyrosine kinases: crossroads in cell signaling and regulation. Trends Endocrinol Metab 17:48–54

Gilbert JA et al (2006) Monoclonal pathogenic antibodies to the thyroid-stimulating hormone receptor in Graves' disease with potent thyroid-stimulating activity but differential blocking activity activate multiple signaling pathways. J Immunol 176:5084–5092

Jacobson DL et al (1997) Epidemiology and estimated population burden of selected autoimmune diseases in the United States. Clin Immunol Immunopath 84:223–243

Johnson KT et al (2013) Examination of orbital tissues in murine models of Graves' disease reveals expression of UCP-1 and the TSHR in retrobulbar adipose tissue. Horm Metab Res 45:401–407

Kaneda T et al (2007) An improved Graves' disease model established by using in vivo electroporation exhibited long-term immunity to hyperthyroidism in BALB/c mice. Endocrinology 148:2335–2344

Kendall-Taylor P, Perros P (1998) Clinical presentation of thyroid associated orbitopathy. Thyroid 8:427–428

Kong YC (1997) Recent developments in the relevance of animal models to Hashimoto's thyroiditis and Graves' disease. Curr Opinion Endocrinol Diab 4:347–353

Krieger CC et al (2015) Bidirectional TSH and IGF-1 receptor cross talk mediates stimulation of hyaluronan secretion by Graves' disease immunoglobulins. J Clin Endocrinol Metabe 100:1071–1077

McLachlan SM et al (2011) Review and hypothesis: does Graves' disease develop in non-human apes. Thyroid 21:1359–1366

McLachlan SM et al (2012) Role of self-tolerance and chronic stimulation in the long term persistence of adenovirus induced thyrotropin receptor antibodies in wild type and transgenic mice. Thyroid 22:931–937

Minich W et al (2013) Autoantibodies to the IGF1 receptor in Graves' orbitopathy. J Clin Endocrinol Metab 98:752–760

Moshkelgosha S et al (2013) Cutting edge: retrobulbar inflammation, adipogenesis, and acute orbital congestion in a preclinical female mouse model of Graves' orbitopathy induced by thyrotropin receptor plasmid-in vivo electroporation. Endocrinology 154:3008–3015

Nagayama Y et al (2002) A novel murine model of Graves' hyperthyroidism with intramuscular injection of adenovirus expressing the thyrotropin receptor. J Immunol 168:2789–2794

Nakahara M et al (2010) Enhanced response to mouse thyroid stimulating hormone (TSH) receptor immunization in TSH receptor knockout mice. Endocrinology 151:4047–4054

Nakahara M et al (2012) Adoptive transfer of antithyrotropin receptor (TSHR) autoimmunity from TSHR knockout mice to athymic nude mice. Endocrinology 153:2034–2042

Pichurin P et al (2001) Naked TSH receptor DNA vaccination: a TH1 T cell response in which interferon-gamma production, rather than antibody, dominates the immune response in mice. Endocrinology 142:3530–3536

Rao P et al (2003) Contrasting activities of thyrotropin receptor antibodies in experimental models of Graves' disease induced by injection of transfected fibroblasts or deoxyribonucleic acid vaccination. Endocrinology 144:260–266

Rapoport B et al (1998) The thyrotropin (TSH) receptor: interactions with TSH and autoantibodies. Endocr Rev 19:673–716

Rose NR (1988) Current concepts of autoimmune disease. Transplant Proc 20(Suppl 4):3–10

Salvi M et al (2014) Efficacy of B-cell targeted therapy with rituximab in patients with active moderate-severe Graves' orbitopathy: a randomized controlled study. J Clin Endocrinol Metabe 100:422–131

Smith RS (2002) Systematic evaluation of the mouse eye: anatomy, pathology, and biomethods. CRC Press, Boca Raton

Stan MN et al (2014) Randomized controlled trial of rituximab in patients with Graves' orbitopathy. J Clin Endocrinol Metabe 100:432–441

Tomer Y, Davies TF (2007) Searching for the autoimmune thyroid disease susceptibility genes: from gene mapping to gene function. Endocrine Rev 24:694–717

Vanderpump MP (2011) The epidemiology of thyroid disease. Br Med Bull 99:39–51

Varewijck AJ et al (2013) Circulating IgGs may modulate IGF-1 receptor stimulating activity in a subset of patients with Graves' ophthalmopathy. J Clin Endocrinol Metab 98:769–776

Villadolid MC et al (1995) Untreated Graves' disease patients without clinical ophthalmopathy demonstrate a high frequency of extraocular muscle (EOM) enlargement by magnetic resonance. J Clin Endocrinol Metab 80:2830–2833

Wall JR (2014) Thyroid function. Pathogenesis of Graves' ophthalmopathy-a role for TSH-R? Nat Rev Endocrinol 10:256–268

Wang SH et al (1998) Induction of thyroiditis in mice with thyrotropin receptor lacking serological dominant regions. Clin Exp Immunol 113:119–125

Weetman AP (2000) Graves' disease. N Engl J Med 343:1236–1248

Weightman DR et al (1993) Autoantibodies to IGF-1 binding sites in thyroid associated ophthalmopathy. Autoimmun 16:251–257

Widera G et al (2000) Increased DNA vaccine delivery and immunogenicity by electroporation in vivo. J Immunol 164:4635–4640

Wiersinga W (2012) Autoimmunity in Graves' ophthalmopathy: the result of an unfortunate marriage between TSH receptors and IGF-1 receptors? J Clin Endocrinol Metab 96:2386–2394

Wiesweg B et al (2013) Current insights into animal models of Graves' disease and orbitopathy. Horm Metab Res 45:549–555

Zhao SX et al (2011) Orbital fibrosis in a mouse model of Graves' disease induced by genetic immunization of thyrotropin receptor cDNA. J Endocrinol 210:369–377

Commentary

Rebecca S. Bahn
e-mail: bahn.rebecca@mayo.edu
Mayo Clinic College of Medicine,
Division of Endocrinology and Metabolism,
Mayo Clinic,
Rochester, MN, USA

Graves' orbitopathy is an autoimmune ocular disease with heterogeneous clinical manifestations. While the majority of patients have a history of hyperthyroidism, some are euthyroid or even hypothyroid. Approximately, 50 % of GO patients have only mild eye irritation and redness. The other half develops more severe disease with forward protrusion of the globes (proptosis), ocular pain and swelling, or debilitating double vision. Some 3–5 % risk loss of sight. The majority experience significantly impaired quality of life. On orbital imaging, most patients show some degree of extraocular muscle enlargement and/or expansion of orbital adipose tissues. The unifying feature of the disease appears to be the presence of circulating autoantibodies directed against the thyrotropin receptor (TSHR) in essentially every patient, given the use of adequately sensitive assays. However, despite the significant morbidity that can accompany the disease, patients with GO and their physicians are left with only limited therapeutic options of having variable effectiveness.

Recent insights into pathogenesis of GO obtained through in vitro studies have undoubtedly moved the field forward. However, the study of potential novel therapy based upon those insights has been severely limited by the lack of an animal model that truly recapitulates the human disease. The robust mouse model developed by Banga, Moshkelgosha, and colleagues represents the first bona fide animal model of GO. The eye changes that develop in these animals are very similar to, and as varied as, those seen in humans with GO. Most mice show extraocular muscle enlargement and some demonstrate expansion of orbital adipose tissues. As in the human disease, all immunized mice produce measurable anti-bodies directed against the TSHR, whether or not they develop hyperthyroidism.

This animal model of GO, and future animal models evolving from it, will facilitate novel experimental approaches and new discoveries regarding GO pathogenesis. However, perhaps its most important contribution to the field will be to offer studies of novel therapies for established disease and approaches to GO prediction and prevention in patients with Graves' hyperthyroidism. These in vitro studies will no doubt lead to randomized clinical trials and ultimately to more effective approaches to care patients with Graves' disease.

Commentary

Shivani Gupta and Raymond S. Douglas
e-mail: raydougl@med.umich.edu
Kellogg Eye Center,
University of Michigan,
Ann Arbor, MI, USA

Graves' disease is a multisystemic autoimmune disorder with a complex underlying etiology, including genetic and environmental factors. Inflammatory infiltration of the orbits is thought to be driven by autoantibodies to the thyrotropin and insulin-like growth factor-1 receptors. Animal models of GO have historically been limited by the inability to simulate the sustained milieu of autoantibodies thought to promote orbital manifestations. The authors describe a mouse model that exhibits a sustained response to the thyrotropin receptor as manifested by induced hyperthyroidism. Close-field electroporation techniques have enabled long lasting autoantibody generation to the thyrotropin receptor and robust antibody response to insulin-like growth factor-1 receptor using in vivo delivery of human purified thyroid stimulating hormone receptor (hTSHR) A-subunit plasmid protein. Histopathological evaluation of orbital tissues in this mouse model reveals the presence of extraocular muscle and perineural inflammatory infiltrates, primarily of CD3[+] T cells. Expansion of orbital fat, chemosis, vascular congestion, and fibrosis at the later stages of disease was analogous to the human disease. Although pathological differences do exist between the animal model of GO and human disease, this model provides considerable advances to further elucidate the complex mechanisms that underlie GO.

Animal Models of Ocular Tumors

9

Martine J. Jager, Jinfeng Cao, Hua Yang, Didier Decaudin, Helen Kalirai, Wietske van der Ent, Nadine E.
de Waard, Nathalie Cassoux, Mary E. Aronow, Rohini
M. Nair and Sarah E. Coupland

9.1 Introduction

For many ocular malignancies, surgery, irradiation, or a combination of both are considered the
first-line treatments. The success rate of primary
therapy is variable among tumor types, but is
usually intermediate to high. However, any subsequent local recurrences or metastases usually
require specific systemic medications, since either surgery or irradiation may not be possible
or effective. The current range of systemic therapeutic agents is limited. In order to broaden the
repertoire of cancer-specific targeted therapies,
pre-clinical in vitro and in vivo testing is necessary. A wide range of animal models are available, for many different types of malignancies,
particularly with the advent of genetic engineering. This chapter will give a summary of the various models used for ocular malignancies, and the
different species that can be used.

We will commence with uveal melanoma
(UM) and its metastases, followed by conjunctival melanoma, retinoblastoma, and, finally, vitreoretinal lymphoma. Much of the information
of this review was obtained from a special edition of "Ocular Oncology and Pathology" (Jager
and Coupland, 2015), in which different animal
models are described more extensively: for more
details, the readers are referred to the respective
articles of this special issue.

M. J. Jager (✉) · J. Cao · N. E. de Waard
Department of Ophthalmology, Leiden University
Medical Center, Albinusdreef 2, 2333 ZA Leiden, The
Netherlands
e-mail: m.j.jager@lumc.nl

H. Yang
Department of Ophthalmology, School of Medicine,
Emory University, Atlanta, Georgia, USA

D. Decaudin
Laboratory of Preclinical Investigation, Translational
Research Department, Institut Curie, Paris, France

H. Kalirai · S. E. Coupland
Pathology, Department of Molecular and Clinical Cancer
Medicine, Institute of Translational Research, University
of Liverpool, Liverpool, UK

W. van der Ent
Institute of Biology, Department of Pathology, Leiden
University Medical Center, Leiden University, Leiden,
The Netherlands

N. Cassoux
Département d'oncologie chirurgicale, Institut and
Laboratory of preclinical investigation, Department of
Translational Research, Institut Curie, Paris, France

M. E. Aronow
Clinical Branch, National Eye Institute,
National Institutes of Health, Bethesda, MD, USA

R. M. Nair
School of Medical Sciences, University of Hyderabad,
Hyderabad, India

9.2 Ocular *Models For Uveal Melanoma*

UM is a rare malignant tumor, occurring in 4–6 new cases per million people (Singh and Topham 2003). Tumors may develop in the iris, the ciliary body, or in the choroid, and systemic dissemination occurs mainly hematogeneously. Up to 50 % of patients may develop metastases, which typically occur in the liver. UM has distinct genetic characteristics, which differ from cutaneous melanoma: early mutations are seen in GNAQ or GNA11 (Van Raamsdonk et al. 2009; Van Raamsdonk et al. 2010). These mutations are mutually exclusive, and experimental work on melanocyte cell lines implanted in mice showed that mutations in these genes play a role in cell division (Van Raamsdonk et al. 2009). This is comparable to mutations in the BRAF gene in cutaneous melanoma. In order for UM metastases to develop, it is proposed that additional genetic alterations are necessary: these include loss of one copy of chromosome 3 and amplification of chromosome 8q (Sisley et al. 1997; Damato et al. 2010; van den Bosch et al. 2012; Cassoux et al. 2014; Versluis et al. 2015). One of the most likely candidate genes for metastatic transformation is located on chromosome 3, namely *BRCA1 associated protein-1*, also termed *BAP1*: loss of mRNA and protein expression due to bi-allelic gene inactivation is highly correlated with the formation of metastases in UM (Harbour et al. 2010; Shah et al. 2013; Koopmans et al. 2014; van Essen et al. 2014; Kalirai et al. 2014).

Experience has demonstrated that UM cell lines characterized by chromosome 3 loss and BAP1 deletion are very difficult to culture. Hence, most of the currently available cell lines have disomy 3 and express BAP1 protein (Griewank et al. 2012). However, recently, BAP1-negative and monosomy 3 cell lines were developed (Amirouchene-Angelozzi et al. 2014). These differing UM cell lines are now being used to test new therapeutic agents, and to determine whether genetic differences between cell lines influence drug susceptibility. Several drugs display a difference in effectiveness, depending on the presence or absence of GNAQ or GNA11 mutations (Chen et al. 2014; Mitsiades et al. 2011), and on the chromosome 3/*BAP1* status (El Filali 2012).

Another characteristic of monosomy 3 UM is an inflammatory phenotype (Maat et al. 2008). Inflammation is recognized as one of the six classic "hallmarks" of cancer (Hanahan and Weinberg 2011), and may play a role in the initiation of tumor growth and in the formation of metastases. UM contains many macrophages, of which the majority are of the pro-angiogenic M2 type (Bronkhorst et al. 2011). The special characteristics of the eye as an immune-privileged site allow growth of tumors in the eye, without the need for immunosuppression. This allows, for instance, studies into the role of the innate and adaptive immune system in murine eye tumors (Ly et al. 2010). However, implantation of human cell lines is often only possible in immunodeficient or immunosuppressed animals.

9.2.1 Spontaneous, Transgenic, and Induced Animal Models

In general, animal models can be essentially divided into three different types: spontaneous, transgenic, and induced models (Cao and Jager 2015). Examples of spontaneous models are intraocular or conjunctival pigmented lesions that arise in some animals, including cats, dogs, and horses. These tumors develop naturally, but are not good models to be used for experimental therapies, as they occur randomly, are rare, and have an unpredictable and more-often-than-not benign clinical course.

Transgenic models offer a great opportunity to study the early stages of malignancies: particular genes are knocked out or knocked in, and subsequently, tumor growth is induced. Most of the known ocular transgenic UM models to date are murine and induce tumors of the retinal pigment epithelium, possibly with the exception of the Tyr-RAS + Ink4a/Arf−/− transgenic mice (Tolleson et al. 2005). However, these murine models do not develop metastases, unless UM cells are first cultured and then transferred into them (Ma and Niederkorn 1995). Non–murine transgenic models include Drosophila and zebrafish. Many biochemical processes are similar in man and Drosophila, and genes can easily be modified to study their effect on cancer processes, for example, the *RB1* gene, or genes that play a role in UM, such as the *GNAQ* and *GNA11* genes and the Polycomb complex, which is related to *BAP1* (reviewed by Bennett et al. 2015). The same can be said about the Zebrafish model, which offers particular advantages highlighted in a review by Van der Ent et al. (Van der Ent et al. 2015).

In induced animal models, tumor development is stimulated by chemical agents, radiation, and xeno-grafted cells or tissues (Cao and Jager 2015). These models are often easier to use for experiments, as they provide reproducible tumors. In UM, the ocular inoculation model is often employed, and tumor cells are injected into the eye, that is, either into the anterior chamber, posterior chamber, or into the subretinal or suprachoroidal spaces. This procedure has been performed on mice, rats, and rabbits (Journee-de Korver et al. 1995; Grossniklaus et al. 1996; Blanco et al. 2000; Rem et al. 2004). While the rabbit model offers the advantage of a large eye, similar in size to that of the human eye, its main disadvantage is the lack of genetically–inbred strains.

9.2.2 Intraocular UM Models

One of the first UM models was developed in the Syrian gold hamster. A cutaneous melanoma was passaged multiple times and developed into a well-growing amelanotic cell line (Greene Melanoma cell line; Greene 1958), which grows successfully inside the eye. Journee-de Korver and coworkers (Journee-de Korver et al. 1992; Journee-de Korver et al. 1995) used tissue implants of this tumor to determine the optimal settings to apply thermotherapy; these settings were subsequently used in UM patients treated with transpupillary thermotherapy. Additionally, this cell

line was used to test the optimal laser settings for transscleral thermotherapy (Rem et al. 2004). The two problems of the Greene melanoma cell line are its limited availability and its skin melanoma origin. An autologous model does exist in the hamster: the Bomirski cutaneous melanoma will grow when a small piece of tissue is transplanted into the hamster eye (Bomirski et al. 1988; Romanowska-Dixon et al. 2001). This tumor model continues to be used extensively in Poland.

Significant work has been undertaken to develop rat ocular xenograft models for UM. Braun and coworkers (Braun et al. 2002; Braun and Vistisen 2012) implanted human melanoma cell lines into the eye of athymic albino WAG/RIjHs-mu rats, either by placing a piece of tissue or minced tumor cells into the choroidal space, or by injecting tumor spheroids into the suprachoroidal space. Most spheroid-injected animals developed eye tumors, which can be used to analyze various UM features, for example, angiogenesis, and to test new drugs. A disadvantage of this model is that it does not develop metastases.

As mentioned above, the rabbit eye is larger than the rat eye and has been used as carrier of several melanoma cell lines. The B16F10 cell line is of murine cutaneous origin and tumor fragments obtained from a mouse skin melanoma will grow when placed in the subchoroidal space of the rabbit eye (Krause et al. 2002). In order for this mouse cell line to successfully grow in the rabbit eye, additional cyclosporin A needs to be administered. In contrast, the Hamster Greene melanoma cell line does not require immunosuppression for growth when placed in a rabbit eye (Romer et al. 1992; Dithmar et al. 2000). A disadvantage of these tumors, however, is their rapid growth, which limits the time for experiments to 10 days (Schuitmaker et al. 1991).

Several human cell lines have also been successfully implanted into rabbit eyes, and occasionally metastases have occurred (Blanco et al. 2000). However, these have tended to spread to the lungs only and did not involve the liver (Blanco et al. 2005). Lopez-Velasco (Lopez-Velasco et al. 2005) tested four different cell lines and noticed that cell lines 92.1 and SP6.5 grow most aggressively. This model has been used quite successfully (Marshall et al. 2007).

9.2.3 UM Metastasis Models

The most frequently used induced model for UM is the mouse eye model. The use of ophthalmological instruments and imaging techniques has enabled development of these UM models, despite the small size of the mouse eye. Some of the murine cell lines give rise to metastases, and also some human cell lines can be used to grow metastases in immunosuppressed mice (reviewed by Yang et al. 2015).

Metastasis models are useful to study the various processes involved in metastasis, that is, tumor cell invasion, migration within the blood stream to the metastatic sites, and the establishment of the metastatic colonies. It is astounding that in human patients, UM metastases can remain dormant for 15–20 years (and in rare cases, even up to 40 years) before becoming clinically manifest, and only then may develop into deadly metastases (Coupland et al. 1996).

As clinical and genetic data of primary UM can be used to stratify patients into high- and low-metastatic risk groups (Damato and Heimann 2013), it is essential that new effective therapeutic agents in metastatic UM are identified. However, testing the effect of drugs on the spreading of primary UM is a problem, as very few of the intraocular melanoma models give rise to metastases within a reasonable experimental time limit. To address this, researchers have injected either human or murine UM cell lines directly into the vasculature or into organs of mice, rats, or rabbits, in order to induce tumor spread, for example, intra-arterially, intracardial injections, injections into the spleen or the liver (reviewed by Cao and Jager 2015 and Yang et al. 2015).

While malignant cells usually grow in the immune-privileged eye, it is unusual for them to disseminate beyond the eye. This is also true for murine and human melanoma cell lines, obtained from either cutaneous or uveal tumors. An excep-

tion can be made for UM cell line 92.1 in the rabbit eye (see above), as this tends to provide metastases quickly. Metastases are not easily found in immunosuppressed or immunodeficient mice, not even when using cell lines derived from human UM metastases, such as OMM2.3. One possibility is that the metastases indeed formed, but the mice were not followed for a sufficiently long period, limiting the usefulness of these models. The formation of metastases depends on the interaction between host and cell lines, in which immunological factors such as HLA expression play a role, and on the characteristics of the cell lines themselves. Niederkorn et al. (Ma et al. 1996; Ma and Niederkorn 1998) implanted several human UM cell lines into the anterior chamber of BALB/c athymic mice. Some of them (OCM1 and 92.1) induced liver metastases reproducibly in a large number of mice. It is intriguing that NK cells in the blood of these mice most likely played a role in the difference of metastasis formation, showing the importance of the immune system in tumor cell dissemination. Heegaard (Heegaard et al. 2003) started to implant UM tissues as xenografts subcutaneously in mice. These developed occasionally, but did not lead to metastases. Surriga (Surriga et al. 2013) labeled the OMM1.3 cell line with enhanced green fluorescent protein (EGFP)-luciferase and implanted the cells retro-orbitally in SCID mice. The use of bioluminescent cells should make regular follow-up easier, as otherwise examination for metastases is only possible when the mouse has been killed.

Experimental treatments of targeted therapies can only be tested on human malignancies, carrying the appropriate mutations. UM cell lines have been used for testing medications (de Lange et al. 2012). Detailed analyses of all available UM cell lines have been undertaken: this is essential to ensure the validity of the cell line as a UM, and to ensure that they have not transformed into another cell line or been inadvertently switched with other cell lines over time (Folberg et al. 2008). There is a need for the development of increased numbers of UM metastatic cell lines, with characteristic mutational profiles of these tumors.

Studies that will provide more insight into the biology of the eye often use murine cell lines, and some of the murine melanoma cell lines will give rise to visible metastases in syngeneic hosts. B16F10 has been used frequently in syngeneic C57Bl6 mice (Niederkorn 1984; Grossniklaus et al. 1995; Ly et al. 2010; de Lange et al. 2012). This cell line has been used extensively to study UM tumor immunology. A derivative cell line of B16 is the Queens melanoma cell line, which frequently metastasizes, B16LS9, and has been used by Grossniklaus to determine the difference between placing cells in the anterior and posterior chamber: anterior chamber inoculation led to fewer metastases than posterior chamber inoculation (Grossniklaus et al. 1996) . Furthermore, this model has been used to study tumor angiogenesis and tumor immunology. Monoclonal antibodies used against human vascular endothelial growth factor (VEGF) affect blood vessels in this mouse model (Yang et al. 2010). All these models use murine skin melanoma cell lines in murine eyes.

9.2.4 Human UM Patient-Derived Xenografts

As mentioned above, injecting UM cells retro-orbitally has recently been used to study the effect of a drug, crizotinib (Surriga et al. 2013). Similar to the experiments of Heegaard, investigators in Paris started to place xenografts derived from primary human UM or from UM metastases in the interscapular fat pad of immunodeficient mice (Garber 2009; Nemati et al. 2010; Decaudin 2011). Tissue obtained from metastases had a better "take" rate than primary tumor material. These grafts were subsequently used to analyze different drugs, alone or in combination. Panels of UM PDXs with different mutations in *GNAQ, GNA11,* and *BAP1* have significant advantages over the cell line-derived xenografts mentioned above, particularly for testing new drugs with the aim of personalized therapy in UM. These PDXs represent the necessary intermediate step between testing drugs in cell lines and their ultimate application in patients.

In order to create a PDX, tumor fragments are obtained from UM patients and directly implanted into immunodeficient mice. The similarity between the PDX and the patients' tumor allows the determination of the sensitivity for clinically available drugs or experimental drugs. At the Curie Institute in Paris, samples from 90 UM or their metastases were obtained, and this led to the development of 25 PDXs growing in severe combined immunodeficiency (SCID) mice. The most successful PDXs were those arising from *GNA11*-mutated UM and their metastases; *BAP1* loss or the UM chromosome 3 status did not appear to affect growth (Nemati et al. 2010; Laurent et al. 2013). The apparent lack of any effect of chromosome 3 may be due to the fact that, of 13 PDXs studied for chromosome status, 10 showed chromosome 3 loss, making further analysis impossible. In vivo passaging did not lead to changes in the main genomic aberrations over time. PDXs showed a complete concordance with the patient material with regard to mutations *(GNA11, GNAQ, BAP1, and SF3B1).* When looking at the gene expression profiles using an Affymetrix Gene Chip, about 3 % of these genes were found to be differentially expressed between the xenograft and the original UM (Laurent et al. 2013). The genes that were overexpressed in the patients were associated with the immune system, extracellular matrix, and angiogenic processes, while the genes that were overexpressed in the PDXs were associated with cell cycle, DNA repair, and kinase activity processes. Early and late passages remained the same (Nemati et al. 2010; Carita et al. 2015).

The PDX can be applied for at least two reasons: to screen potential drugs to be used for UM, and to test a drug for a particular patient. It may be possible to initially determine the mutation or genome status of a particular UM, perform an expression array, or determine the presence of specific kinases, and then add the most probable therapeutic agent(s) and study the effect in the mouse PDXs. This model can be applied to test different combinations of drugs. Although this is a very costly procedure to be used in every case, it may ultimately lead to "personalized" medicine in the therapy of metastatic UM, gaining rapid insight into the sensitivity of UM metastases to a specific drug, prior to their ultimate application in patients.

9.2.5 Zebrafish UM Model

While melanoma cells will grow successfully when they are engrafted in mice, tumor cells may also grow in zebrafish *(Danio rerio)*. The embryonic zebrafish can be used as a carrier of UM cells. These cells can be injected into the embryo, and since the adaptive immune system has not yet been developed in the embryo, immunosuppressive drugs are not needed. UM cells can be injected into the yolk sack, the cardiac cavity, or directly into blood vessels, and the ability of the tumor cells to spread can be studied.

The advantages of a zebrafish model include the following: (a) many animals can be obtained and studied simultaneously; (b) noninvasive live imaging is possible due to physical transparency of the developing embryo; and (c) the zebrafish does not have a functional adaptive immune system at early stages of development. This means that immunosuppressive drugs are not required, nor that the experiments are limited to specific immunodeficient strains. Furthermore, the availability of various fluorescent reporter lines aids in investigations by examining the interactions between tumor cells and host. After injection of tumor cells into zebrafish, cell migration, proliferation, and formation of metastases can be studied.

Injection of UM cells into the yolk sack indeed leads to proliferation and migration of UM cells. Injection of biofluorescent cells helps to follow the migration route of the cells and quantify tumor cell burden in vivo. The effect of drugs on all of these processes can be evaluated, and one only has to add the investigational drug to the water in which the fish are maintained. Van der Ent et al. (2014, 2015) tried several injection routes to develop an animal model. Tumor cells that express mCherry were injected into the yolk sac of 2 days post-fertilization (dpf) embryos.

Live confocal imaging was performed on embryos at 4, 5, and 6 days post injection (dpi) and the route of migration was analyzed: at 4 dpi, cells started to migrate out of the yolk sac. Migration occurred both to the tail and head region, and also along blood vessels. Cells from the metastasis-derived tumor cell line OMM2.3 showed more migration than of the primary tumor-derived cell lines Mel270 and 92.1. Using this model, it was possible to show the effect of several drugs (Van der Ent et al. 2014, 2015). Different effects were observed with regard to, for instance, crizotinib, a c-Met inhibitor, which limited migration of c-Met positive OMM2.3 cells, but had hardly any effect on the c-Met low 92.1 cells.

An intriguing finding is that after yolk sac injections, about 10% of the embryos showed the presence of UM cells in one or both eyes of the embryo (Van der Ent et al. 2015).

Another route of tumor cell injection is directly into the blood stream of 2 dpf embryos. Cells can be injected into the heart cavity or into the Duct of Cuvier. When using this technique, 80% of embryos develop intraocular tumor growth. The zebrafish model has been used to grow freshly obtained tumor material as a xenograft (Marques et al. 2009).

9.2.6 Chick Embryo UM Model

A further way to examine UM metastasis is by utilization of the chick embryo (reviewed by Kalirai et al. 2015). Tumor cells can be placed on the highly vascularized chorioallantoic membrane (CAM), where a tumor mass will then develop. Tumor cells invade into the CAM's blood vessels. A second approach is through the injection of tumor cells directly into the blood stream.

The CAM begins to develop by day 4 and by day 12, consists of three layers: the ectoderm, the mesoderm that contains blood vessels, and the endoderm. The ectoderm has a capillary plexus. Additionally, lymph vessels start to develop. The immune system is not yet developed at this time. The chick embryo has been used for xenografting of tumor cells for over a century (Rous and

Murphy 1912), but studies on UM are from recent years. Luyten used this model to inject UM cells directly into the eye (Luyten et al. 1993), where tumor growth was observed.

Kalirai et al. (2015) used the CAM assay: space is made above the CAM inside the egg by removing some albumen. A small opening is made in the shell of the egg, and UM cells can be placed on the CAM through this window. Cell lines OMM1 and 92.1 were engrafted to an area with blood vessels on day 7, and after 1 week, a tumor nodule had developed. These cell lines showed a high Ki-67 proliferation index, Melan-A expression, and vascular structures. Berube (Berube et al. 2005) achieved a similar result with four other UM cell lines.

Green fluorescent protein (GFP)-labeled OMM1 and 92.1 cells were also directly injected into the circulation (Luyten et al. 1996; De Waard-Siebinga et al. 1995; Kalirai et al. 2015). Cells specifically homed to the neural crest-derived tissues, including the uveal tract of the eye and the liver (Kalirai et al. 2015) . This intravascular model can be used to examine the differences between genetically–different UM cell lines, modified cell lines, or the effect of drugs on tumor growth. The CAM assay is also appropriate for this. One can follow the growth of fluorescent cells in vivo through intravital video-microscopy, which has been used to analyze the B16F1 melanoma development in chick embryos (Chambers et al. 1992).

9.3 Other Ocular Tumor Models

9.3.1 Conjunctival Melanoma

Conjunctival melanoma is even rarer than its intraocular "cousin," and is often associated with many clinical problems, for example, tumors are often multifocal, they often recur locally, they may lack pigmentation, and there are various premalignant noninvasive stages making it difficult to detect and diagnose this disease (Missotten et al. 2005; Shields et al. 2011; Kenawy et al. 2013). The mutational profile of

conjunctival melanoma differs from that of UM and is characterized by *BRAF* and *NRAS* mutations in up to 50 and 20 % of cases, respectively (Gear et al. 2004; Spendlove et al. 2004; Lake et al. 2011; Griewank et al. 2013; Weber et al. 2013). Conjunctival melanoma spreads typically via the lymphatics, and at present there are few effective drugs for treatment of these metastases. It is therefore of great importance to develop better therapies, and in order to test drugs for this malignancy, in vitro and in vivo models are required.

While cell culture is often used for testing medications, the availability of an animal model for conjunctival melanoma would be very useful. De Waard recently developed a conjunctival melanoma metastases model (De Waard et al. 2015a, b). She used three different conjunctival cell lines, of which two have a *BRAF* (T1799A) mutation (cell lines CRMM-1 and CM2005.1), and one cell line, CRMM-2, an *NRAS* (A182T) mutation (Nareyeck et al. 2005; Keijser et al. 2007; Gear et al. 2004; Griewank et al. 2013; De Waard et al. 2015a, b). In vitro grown cells were injected subconjunctivally into genetically–immunodeficient NOD/SCID interleukin (IL)-2 ry^{null} mice. The three cell lines led to different in vivo characteristics: the primary tumors of CRMM-1 lacked pigment, those of CRMM-2 had some pigment, while the primary tumors of CM2005.1 were heavily pigmented. Although these subconjunctival injections led to tumor growth in all cases, no metastases were seen. Two cell lines grew quite rapidly and when single-cell suspensions were obtained from their subconjunctival xenografts and subsequently implanted subconjunctivally into a second group of mice, metastases ultimately developed.

This conjunctival melanoma metastasis model was used to determine whether a specific stem cell marker is expressed in conjunctival melanoma cell lines (De Waard et al. 2015a, b). ABCB5 is present on cutaneous melanoma cells, where it identifies a subpopulation of cells that play a role in tumor recurrence and show an increased tumorigenicity (Schatton et al. 2008; Wilson et al. 2011). This marker may therefore constitute a good therapeutic target, and its presence was determined in the cell lines and the different tumor stages in the murine model. A minority of cells within all three cell lines expressed the stem cell marker ABCB5, and this marker was also expressed in the xenografts. Growing tumors and metastases had a higher expression of this marker. It will be interesting to use this model to test the effect of biologics directed against ABCB5.

9.3.2   Retinoblastoma

Retinoblastoma is a malignancy that occurs in children, and is much more prevalent than UM in many parts of the world. The disease develops when both alleles of the retinoblastoma gene, *RB1*, produce a nonfunctional protein (Theriault et al. 2014). The disease can occur in one eye or in both eyes, in the latter case most often when the patient inherited one defective copy of *RB1*. In such a case, one may try to save at least one eye. Originally, irradiation was the treatment of choice, but this may lead to deformities and secondary cancers. Different other treatments are available, but many modalities use chemotherapeutic agents. These can be injected intravenously, intra-arterially, subconjunctivally, or into the vitreous (Shields and Shields 2004).

As in UM, cell lines and animal models are being used to further improve therapies in retinoblastoma. With respect to retinoblastoma cell lines, only a few drugs are being used experimentally, and the results of these tests cannot necessarily be extrapolated to all tumors. The available cell lines Y79 and WERI-Rb have been used in animal experiments: they will grow in the anterior chamber of immunosuppressed mice (Benedict et al. 1980; Aerts et al. 2010), rabbits (McFall et al. 1977), in the subretinal space (Kang and Grossniklaus 2011), or they can be injected into the vitreous cavity of immunodeficient mice (Chevez-Barrios et al. 2000) or rats (Bibby 2004; Laurie et al. 2005).

Cassoux et al. (2015) are developing a new model of retinoblastoma xenografts, in which new drugs can be tested on patient material. The first xenografts were placed subcutaneously in immunodeficient mice and used for drug testing. Subsequently, an orthotopic model was also developed. The subcutaneously–growing tumors from the nude mice were suspended and injected into the subretinal space. The development of the subretinal tumors was followed by optical coherence tomography (OCT). Tumors grew into the vitreous. This model was used to test the efficacy of intravitreal injections of melphalan, carboplatina, and bevacizumab. By using this type of models, new drugs can thus be properly tested prior to human use and new application routes can be investigated.

9.3.3   Transgenic Mouse Models of Retinoblastoma

Although transgenic models are not commonly used for melanocytic eye tumors, they have been used extensively for retinoblastoma. The LHβTag mouse was developed through the use of LHβ gene as promotor and simian virus 40 (SV40) large T antigen, and mice with this transgene developed bilateral ocular tumors, due to the presence of several oncogenes (Windle et al. 1990). The intraocular tumors share many characteristics with retinoblastoma, with both endophytic and exophytic growth, optic nerve invasion, and vitreous seeding. The Tag model has been used to study tumor development and to test potential therapies. A commonly–applied model utilizes the promotor of the interphotoreceptor retinoid-binding protein together with Tag (Al-Ubaidi et al. 1992). The problem with this model is that the mice do not develop metastases and do not exhibit the extensive apoptosis seen in human retinoblastoma.

Another approach has been the development of knockout mice, especially when *Rb* and *p107* were both inactivated. The *p107* gene was found to be necessary to facilitate the oncogenic program following Rb inactivation (Robanus-Maandag et al. 1998). Rb/p107 and Rb/p130, double knockout models, develop retinal tumors (reviewed by Nair and Vemuganti 2015). The great advantage of these models is the similar behavior in different mice, thus allowing testing of a variety of drugs. Furthermore, one can study the role of cancer stem cells, angiogenesis, the role of macrophages, the formation of metastases, and therapeutic resistance, among others.

9.3.4 Vitreoretinal Lymphoma

Primary vitreoretinal lymphoma (VRL) is a malignant disease of the nervous system, mainly of B-cell origin (Coupland and Damato 2008; Chan et al. 2011). It is a high-grade B-cell non-Hodgkin lymphoma and the malignant cells can infiltrate the eye as well as the brain. While about 25% of patients with primary central nervous system lymphoma (PCNSL) have eye involvement, more than 50% of VRL spread to the brain. An important discussion is understanding how VRL cells move between the brain and the eye: this is a feature that can be examined in an animal model.

Animal models aid lymphoma cell growth, as the malignant lymphocytes show a lack of proliferative capacity in vitro (reviewed by Aronow et al. 2015). White and coworkers (White et al. 1984) obtained T cell blasts from the blood of a child, and injected these into the anterior chambers of nude mice. Cells were maintained through intraocular transfers (White et al. 1990). B-cell lymphoma models have also been developed for VRL (Li et al. 2006).

Assaf et al. (1997) demonstrated that cells may migrate from the eye through the choroid plexus and cranial nerves (Hochberg and Miller 1988). Chan (Chan et al. 2005) used a lymphoma cell line, Rev-2-T-6, to create a vitreous model in BALB/c mice. The finding that this resulted in high levels of IL-10 and interferon (IFN)-gamma inside the eye was subsequently used to help the differentiation between uveitis and lymphoma in patients. The disadvantage of this model was that it utilized T cells, but, similarly, B-cell lymphoma cells have been injected into the eyes of immunodeficient mice. These models were used to show the efficacy of using B-cell-specific monoclonal antibodies for the treatment of B-cell lympho-mas (Bang et al. 2005; Smith et al. 2007), also for intracerebral locations (Mineo et al. 2008). A similar effective therapy was obtained in a mixed ocular lymphoma/central nervous system B-cell lymphoma model using an anti-human CD20 monoclonal antibody (Ben Abdelwahed et al. 2013). Anti-B-cell monoclonal antibodies are now widely being used clinically.

9.4 Conclusion

Significant progress has been made in the field of ocular oncology therapeutics for the treatment of primary tumors. Despite this, targeted therapies are urgently required when metastatic spread of ocular tumors occurs, for example, metastatic uveal melanoma, metastatic retinoblastoma, and cerebral involvement of vitreoretinal lymphoma. In such cases, the response to currently available chemotherapy or immunotherapy is minimal and, therefore, the prognosis is generally poor. Preclinical models—cell lines, cell cultures, cell line-derived xenografts and PDXs—can help investigate the factors involved in malignant transformation, invasion, and metastasis as well as to examine the response to therapy of ocular tumors. Each of the models has its limitations and each can only address part of the biological question we pose; hence we must ensure that we interpret the data within the framework of the limitations of the assay used. Despite this, these animal models can potentially provide an intermediary step and the preliminary data required prior to the introduction of therapeutics in the form of early clinical trials. Both new technologies in genetic animal modeling and in targeted drug development offer exciting opportunities in developing effective therapies, which hopefully will improve the outcomes of disseminated ocular malignancies.

9.4.1 Compliance with Ethical Requirements

Martine J Jager, Jinfeng Cao, Hua Yang, Didier Decaudin, Helen Kalirai, Wietske van der Ent,

Nadine de Waard, Nathalie Cassoux, Mary E. Aronow, Rohini M. Nair, Sarah E Coupland, and Arun Singh declare that they have no conflict of interest. All institutional and national guidelines for the care and use of laboratory animals were followed.

References

Aerts I, Leuraud P, Blais J, Pouliquen AL, Maillard P, Houdayer C, Couturier J, Sastre-Garau X, Grierson D, Doz F, Poupon MF (2010) In vivo efficacy of photodynamic therapy in three new xenograft models of human retinoblastoma. Photodiagnosis Photodyn Ther 7(4):275–283. doi:10.1016/j.pdpdt.2010.09.003

Al-Ubaidi MR, Font RL, Quiambao AB, Keener MJ, Liou GI, Overbeek PA, Baehr W (1992) Bilateral retinal and brain tumors in transgenic mice expressing simian virus 40 large T antigen under control of the human interphotoreceptor retinoid-binding protein promoter. J Cell Biol 119(6):1681–1687

Amirouchene-Angelozzi N, Nemati F, Gentien D, Nicolas A, Dumont A, Carita G, Camonis J, Desjardins L, Cassoux N, Piperno-Neumann S, Mariani P, Sastre X, Decaudin D, Roman-Roman S (2014) Establishment of novel cell lines recapitulating the genetic landscape of uveal melanoma and preclinical validation of mTOR as a therapeutic target. Mol Oncol 8(8):1508–1520. doi:10.1016/j.molonc.2014.06.004

Aronow ME, Shen D, Hochman J, Chan CC (2015) Intraocular lymphoma models. Ocular Oncol Pathol 1:214–222. doi:10.1159/000370158

Assaf N, Hasson T, Hoch-Marchaim H, Pe'er J, Gnessin H, Deckert-Schluter M, Wiestler OD, Hochman J (1997) An experimental model for infiltration of malignant lymphoma to the eye and brain. Virchows Arch 431(6):459–467

Bang S, Nagata S, Onda M, Kreitman RJ, Pastan I (2005) HA22 (R490A) is a recombinant immunotoxin with increased antitumor activity without an increase in animal toxicity. Clin Cancer Res 11(4):1545–1550. doi:10.1158/1078-0432.CCR-04-1939

Ben Abdelwahed R, Donnou S, Ouakrim H, Crozet L, Cosette J, Jacquet A, Tourais I, Fournes B, Gillard Bocquet M, Miloudi A, Touitou V, Daussy C, Naud MC, Fridman WH, Sautes-Fridman C, Urbain R, Fisson S (2013) Preclinical study of Ublituximab, a glycoengineered anti-human CD20 antibody, in murine models of primary cerebral and intraocular B-cell lymphomas. Invest Ophthalmol Vis Sci 54(5):3657–3665. doi:10.1167/iovs.12-10316

Benedict WF, Dawson JA, Banerjee A, Murphree AL (1980) The nude mouse model for human retinoblastoma: a system for evaluation of retinoblastoma therapy. Med Pediatr Oncol 8(4):391–395

Bennett D, Lyulcheva E, Cobbe N (2015) Drosophila as a potential model for ocular tumors. Ocular Oncol Pathol 1:190–199.

Berube M, Deschambeault A, Boucher M, Germain L, Petitclerc E, Guerin SL (2005) MMP-2 expression in uveal melanoma: differential activation status dictated by the cellular environment. Mol Vis 11:1101–1111

Bibby MC (2004) Orthotopic models of cancer for preclinical drug evaluation: advantages and disadvantages. Eur J Cancer 40(6):852–857. doi:10.1016/j.ejca.2003.11.021

Blanco G, Saornil AM, Domingo E, Diebold Y, Lopez R, Rabano G, Tutor CJ (2000) Uveal melanoma model with metastasis in rabbits: effects of different doses of cyclosporine A. Curr Eye Res 21(3):740–747

Blanco PL, Marshall JC, Antecka E, Callejo SA, Souza Filho JP, Saraiva V, Burnier MN, Jr. (2005) Characterization of ocular and metastatic uveal melanoma in an animal model. Invest Ophthalmol Vis Sci 46(12):4376–4382. doi:10.1167/iovs.04-1103

Bomirski A, Slominski A, Bigda J (1988) The natural history of a family of transplantable melanomas in hamsters. Cancer Metastasis Rev 7(2):95–118

Braun RD, Vistisen KS (2012) Modeling human choroidal melanoma xenograft growth in immunocompromised rodents to assess treatment efficacy. Invest Ophthalmol Vis Sci 53(6):2693–2701. doi:10.1167/iovs.11-9265

Braun RD, Abbas A, Bukhari SO, Wilson W, 3rd (2002) Hemodynamic parameters in blood vessels in choroidal melanoma xenografts and rat choroid. Invest Ophthalmol Vis Sci 43(9):3045–3052

Bronkhorst IH, Ly LV, Jordanova ES, Vrolijk J, Versluis M, Luyten GP, Jager MJ (2011) Detection of M2-macrophages in uveal melanoma and relation with survival. Invest Ophthalmol Vis Sci 52(2):643–650. doi:10.1167/iovs.10-5979

Cao J, Jager MJ (2015) Animal eye models for uveal melanoma. Ocular Oncol Pathol 1:141–150. doi:10.1159/000370152

Carita G, Nemati F, Decaudin D (2015) Uveal melanoma patient-derived xenografts. Ocular Oncol Pathol 1:161–169. doi:10.1159/000370154

Cassoux N, Rodrigues MJ, Plancher C, Asselain B, Levy-Gabriel C, Lumbroso-Le Rouic L, Piperno-Neumann S, Dendale R, Sastre X, Desjardins L, Couturier J (2014) Genome-wide profiling is a clinically relevant and affordable prognostic test in posterior uveal melanoma. Br J Ophthalmol 98(6):769–774. doi:10.1136/bjophthalmol-2013-303867

Cassoux N, Thuleau A, Assayag F, Aerts I, Decaudin D (2015) Establishment of an orthotopic xenograft mice model of retinoblastoma suitable for preclinical testing. Ocular Oncol Pathol 1:200–206. doi:10.1159/000370156

Chambers AF, Schmidt EE, MacDonald IC, Morris VL, Groom AC (1992) Early steps in hematogenous metastasis of B16F1 melanoma cells in chick embryos studied by high-resolution intravital videomicroscopy. J Natl Cancer Inst 84(10):797–803

Chan CC, Fischette M, Shen D, Mahesh SP, Nussenblatt RB, Hochman J (2005) Murine model of primary intraocular lymphoma. Invest Ophthalmol Vis Sci 46(2):415–419. doi:10.1167/iovs.04-0869

Chan CC, Rubenstein JL, Coupland SE, Davis JL, Harbour JW, Johnston PB, Cassoux N, Touitou V, Smith JR, Batchelor TT, Pulido JS (2011) Primary vitreoretinal lymphoma: a report from an International Primary Central Nervous System Lymphoma Collaborative Group symposium. Oncologist 16(11):1589–1599. doi:10.1634/theoncologist.2011-0210

Chen X, Wu Q, Tan L, Porter D, Jager MJ, Emery C, Bastian BC (2014) Combined PKC and MEK inhibition in uveal melanoma with GNAQ and GNA11 mutations. Oncogene 33(39):4724–4734. doi:10.1038/onc.2013.418

Chevez-Barrios P, Hurwitz MY, Louie K, Marcus KT, Holcombe VN, Schafer P, Aguilar-Cordova CE, Hurwitz RL (2000) Metastatic and nonmetastatic models of retinoblastoma. Am J Pathol 157(4):1405–1412. doi:10.1016/S0002-9440(10)64653-6

Coupland SE, Damato B (2008) Understanding intraocular lymphomas. Clin Experiment Ophthalmol 36(6):564–578. doi:10.1111/j.1442-9071.2008.01843.x

Coupland SE, Sidiki S, Clark BJ, McClaren K, Kyle P, Lee WR (1996) Metastatic choroidal melanoma to the contralateral orbit 40 years after enucleation. Arch Ophthalmol 114(6):751–756

Damato B, Heimann H (2013) Personalized treatment of uveal melanoma. Eye 27(2):172–179

Damato B, Dopierala JA, Coupland SE (2010) Genotypic profiling of 452 choroidal melanomas with multiplex ligation-dependent probe amplification. Clin Cancer Res 16(24):6083–6092. doi:10.1158/1078-0432.CCR-10-2076

De Lange J, Ly LV, Lodder K, Verlaan-de Vries M, Teunisse AF, Jager MJ, Jochemsen AG (2012) Synergistic growth inhibition based on small-molecule p53 activation as treatment for intraocular melanoma. Oncogene 31(9):1105–1116. doi:10.1038/onc.2011.309

De Waard-Siebinga I, Blom DJ, Griffioen M, Schrier PI, Hoogendoorn E, Beverstock G, Danen EH, Jager MJ (1995) Establishment and characterization of an uveal-melanoma cell line. Int J Cancer 62(2):155–161

De Waard NE, Kolovou PE, McGuire SP, Cao J, Frank NY, Frank MH, Jager MJ, Ksander BR (2015a) Expression of multidrug resistance transporter ABCB5 in a murine model of human conjunctival melanoma. Ocular Oncol Pathol 1:182–189. doi: 10.1159/000371555.

De Waard NE, Cao J, McGuire SP, Kolovou PE, Jordanova ES, Ksander BR, Jager MJ (2015b) A murine model for metastatic conjunctival melanoma. Invest Ophthalmol Vis Sci doi:10.1167/iovs.14-15239

Decaudin D (2011) Primary human tumor xenografted models ("tumorgrafts") for good management of patients with cancer. Anticancer Drugs 22(9):827–841. doi:10.1097/CAD.0b013e3283475f70

Dithmar S, Albert DM, Grossniklaus HE (2000) Animal models of uveal melanoma. Melanoma Res 10(3):195–211

El Filali M (2012) Knowledge-based treatment in uveal melanoma. Thesis Universiteit Leiden 2012.

Folberg R, Kadkol SS, Frenkel S, Valyi-Nagy K, Jager MJ, Pe'er J, Maniotis AJ (2008) Authenticating cell lines in ophthalmic research laboratories. Invest Ophthalmol Vis Sci 49(11):4697–4701. doi:10.1167/iovs.08-2324

Garber K (2009) From human to mouse and back: "tumorgraft" models surge in popularity. J Natl Cancer Inst 101(1):6–8. doi:10.1093/jnci/djn481

Gear H, Williams H, Kemp EG, Roberts F (2004) BRAF mutations in conjunctival melanoma. Invest Ophthalmol Vis Sci 45(8):2484–2488. doi:10.1167/iovs.04-0093

Greene HS (1958) A spontaneous melanoma in the hamster with a propensity for amelanotic alteration and sarcomatous transformation during transplantation. Cancer Res 18(4):422–425

Griewank KG, Yu X, Khalili J, Sozen MM, Stempke-Hale K, Bernatchez C, Wardell S, Bastian BC, Woodman SE (2012) Genetic and molecular characterization of uveal melanoma cell lines. Pigment Cell Melanoma Res 25(2):182–187. doi:10.1111/j.1755-148X.2012.00971.x

Griewank KG, Westekemper H, Murali R, Mach M, Schilling B, Wiesner T, Schimming T, Livingstone E, Sucker A, Grabellus F, Metz C, Susskind D, Hillen U, Speicher MR, Woodman SE, Steuhl KP, Schadendorf D (2013) Conjunctival melanomas harbor BRAF and NRAS mutations and copy number changes similar to cutaneous and mucosal melanomas. Clin Cancer Res 19(12):3143–3152. doi:10.1158/1078-0432.CCR-13-0163

Grossniklaus HE, Barron BC, Wilson MW (1995) Murine model of anterior and posterior ocular melanoma. Curr Eye Res 14(5):399–404

Grossniklaus HE, Wilson MW, Barron BC, Lynn MJ (1996) Anterior vs posterior intraocular melanoma. Metastatic differences in a murine model. Arch Ophthalmol 114(9):1116–1120

Hanahan D, Weinberg RA (2011) Hallmarks of cancer: the next generation. Cell 144(5):646–674. doi:10.1016/j.cell.2011.02.013

Harbour JW, Onken MD, Roberson ED, Duan S, Cao L, Worley LA, Council ML, Matatall KA, Helms C, Bowcock AM (2010) Frequent mutation of BAP1 in metastasizing uveal melanomas. Science 330(6009):1410–1413. doi:10.1126/science.1194472

Heegaard S, Spang-Thomsen M, Prause JU (2003) Establishment and characterization of human uveal malignant melanoma xenografts in nude mice. Melanoma Res 13(3):247–251. doi:10.1097/01.cmr.0000056239.78713.c8

Hochberg FH, Miller DC (1988) Primary central nervous system lymphoma. J Neurosurg 68(6):835–853. doi:10.3171/jns.1988.68.6.0835

Jager MJ, Coupland SE (editors) (2015) Animal models in ocular oncology. Ocular Oncol Pathol 1(3):123–222

Journee-de Korver JG, Oosterhuis JA, Kakebeeke-Kemme HM, de Wolff-Rouendaal D (1992) Transpupillary thermotherapy (TTT) by infrared irradiation of choroidal melanoma. Doc Ophthalmol 82(3):185–191

Journee-de Korver JG, Oosterhuis JA, Vrensen GF (1995) Light and electron microscopic findings on experimental melanomas after hyperthermia at 50 degrees C. Melanoma Res 5(6):393–402

Kalirai H, Dodson A, Faqir S, Damato B, Coupland SE (2014) Lack of BAP1 protein expression in uveal melanoma is associated with increased metastatic risk and has utility in routine prognostic testing. Br J Cancer 111(7):1373–1380

Kalirai H, Shahidipour H, Coupland SE, Luyten GPM (2015) Use of the chick embryo model in uveal melanoma. Ocular Oncol Pathol 1:133–140. doi:10.1159/000370151

Kang SJ, Grossniklaus HE (2011) Rabbit model of retinoblastoma. J Biomed Biotechnol 2011:394730. doi:10.1155/2011/394730

Keijser S, Maat W, Missotten GS, de Keizer RJ (2007) A new cell line from a recurrent conjunctival melanoma. Br J Ophthalmol 91(11):1566–1567. doi:10.1136/bjo.2006.110841

Kenawy N, Lake SL, Coupland SE, Damato BE (2013) Conjunctival melanoma and melanocytic intra-epithelial neoplasia. Eye 27(2):142–152

Koopmans AE, Verdijk RM, Brouwer RW, van den Bosch TP, van den Berg MM, Vaarwater J, Kockx CE, Paridaens D, Naus NC, Nellist M, van IWF, Kilic E, de Klein A (2014) Clinical significance of immunohistochemistry for detection of BAP1 mutations in uveal melanoma. Mod Pathol 27(10):1321–1330. doi:10.1038/modpathol.2014.43

Krause M, Kwong KK, Xiong J, Gragoudas ES, Young LH (2002) MRI of blood volume and cellular uptake of superparamagnetic iron in an animal model of choroidal melanoma. Ophthalmic Res 34(4):241–250. doi:63883

Lake SL, Jmor F, Dopierala J, Taktak AF, Coupland SE, Damato BE (2011) Multiplex ligation-dependent probe amplification of conjunctival melanoma reveals common BRAF V600E gene mutation and gene copy number changes. Invest Ophthalmol Vis Sci 52(8):5598–5604. doi:10.1167/iovs.10-6934

Laurent C, Gentien D, Piperno-Neumann S, Nemati F, Nicolas A, Tesson B, Desjardins L, Mariani P, Rapinat A, Sastre-Garau X, Couturier J, Hupe P, de Koning L, Dubois T, Roman-Roman S, Stern MH, Barillot E, Harbour JW, Saule S, Decaudin D (2013) Patient-derived xenografts recapitulate molecular features of human uveal melanomas. Mol Oncol 7(3):625–636. doi:10.1016/j.molonc.2013.02.004

Laurie NA, Gray JK, Zhang J, Leggas M, Relling M, Egorin M, Stewart C, Dyer MA (2005) Topotecan combination chemotherapy in two new rodent models of retinoblastoma. Clin Cancer Res 11(20):7569–7578. doi:10.1158/1078-0432.CCR-05-0849

Li Z, Mahesh SP, Shen de F, Liu B, Siu WO, Hwang FS, Wang QC, Chan CC, Pastan I, Nussenblatt RB (2006) Eradication of tumor colonization and invasion by a B cell-specific immunotoxin in a murine model for human primary intraocular lymphoma. Cancer Res 66(21):10586–10593. doi:10.1158/0008-5472.CAN-06-1981

Lopez-Velasco R, Morilla-Grasa A, Saornil-Alvarez MA, Ordonez JL, Blanco G, Rabano G, Fernandez N, Almaraz A (2005) Efficacy of five human melanocytic cell lines in experimental rabbit choroidal melanoma. Melanoma res 15(1):29–37

Luyten GP, Mooy CM, De Jong PT, Hoogeveen AT, Luider TM (1993) A chicken embryo model to study the growth of human uveal melanoma. Biochem Biophys Res Commun 192(1):22–29

Luyten GP, Naus NC, Mooy CM, Hagemeijer A, Kan-Mitchell J, Van Drunen E, Vuzevski V, De Jong PT, Luider TM (1996) Establishment and characterization of primary and metastatic uveal melanoma cell lines. Int J Cancer 66(3):380–387. doi:10.1002/(SICI)1097-0215(19960503)66:3<380::AID-IJC19>3.0.CO;2-F

Ly LV, Baghat A, Versluis M, Jordanova ES, Luyten GP, van Rooijen N, van Hall T, van der Velden PA, Jager MJ (2010) In aged mice, outgrowth of intraocular melanoma depends on proangiogenic M2-type macrophages. J Immunol 185(6):3481–3488. doi:10.4049/jimmunol.0903479

Ma D, Niederkorn JY (1995) Efficacy of tumor-infiltrating lymphocytes in the treatment of hepatic metastases arising from transgenic intraocular tumors in mice. Invest Ophthalmol Vis Sci 36(6):1067–1075

Ma D, Niederkorn JY (1998) Role of epidermal growth factor receptor in the metastasis of intraocular melanomas. Invest Ophthalmol Vis Sci 39 (7):1067–1075

Ma D, Luyten GP, Luider TM, Jager MJ, Niederkorn JY (1996) Association between NM23-H1 gene expression and metastasis of human uveal melanoma in an animal model. Invest Ophthalmol Vis Sci 37(11):2293–2301

Maat W, Ly LV, Jordanova ES, de Wolff-Rouendaal D, Schalij-Delfos NE, Jager MJ (2008) Monosomy of chromosome 3 and an inflammatory phenotype occur together in uveal melanoma. Invest Ophthalmol Vis Sci 49(2):505–510. doi:10.1167/iovs.07-0786

Marques IJ, Weiss FU, Vlecken DH, Nitsche C, Bakkers J, Lagendijk AK, Partecke LI, Heidecke CD, Lerch MM, Bagowski CP (2009) Metastatic behaviour of primary human tumours in a zebrafish xenotransplantation model. BMC Cancer 9:128. doi:10.1186/1471-2407-9-128

Marshall JC, Fernandes BF, Di Cesare S, Maloney SC, Logan PT, Antecka E, Burnier MN, Jr. (2007) The use of a cyclooxygenase-2 inhibitor (Nepafenac) in an ocular and metastatic animal model of uveal melanoma. Carcinogenesis 28(9):2053–2058. doi:10.1093/carcin/bgm091

McFall RC, Sery TW, Makadon M (1977) Characterization of a new continuous cell line derived from a human retinoblastoma. Cancer Res 37(4):1003–1010

Mineo JF, Scheffer A, Karkoutly C, Nouvel L, Kerdraon O, Trauet J, Bordron A, Dessaint JP, Labalette M, Berthou C, Labalette P (2008) Using human CD20-transfected murine lymphomatous B cells to evaluate the efficacy of intravitreal and intracerebral rituximab injections in mice. Invest Ophthalmol Vis Sci 49(11):4738–4745. doi:10.1167/iovs.07-1494

Missotten GS, Keijser S, De Keizer RJ, De Wolff-Rouendaal D (2005) Conjunctival melanoma in the Netherlands: a nationwide study. Invest Ophthalmol Vis Sci 46(1):75–82. doi:10.1167/iovs.04-0344

Mitsiades N, Chew SA, He B, Riechardt AI, Karadedou T, Kotoula V, Poulaki V (2011) Genotype-dependent sensitivity of uveal melanoma cell lines to inhibition of B-Raf, MEK, and Akt kinases: rationale for personalized therapy. Invest Ophthalmol Vis Sci 52(10):7248–7255. doi:10.1167/iovs.11-7398

Nair RM, Vemuganti GK (2015) Transgenic models in retinoblastoma research. Ocular Oncol Pathol 1:207–213. doi:10.1159/000370157

Nareyeck G, Wuestemeyer H, von der Haar D, Anastassiou G (2005) Establishment of two cell lines derived from conjunctival melanomas. Exp Eye Res 81(3):361–362. doi:10.1016/j.exer.2005.04.018

Nemati F, Sastre-Garau X, Laurent C, Couturier J, Mariani P, Desjardins L, Piperno-Neumann S, Lantz O, Asselain B, Plancher C, Robert D, Peguillet I, Donnadieu MH, Dahmani A, Bessard MA, Gentien D, Reyes C, Saule S, Barillot E, Roman-Roman S, Decaudin D (2010) Establishment and characterization of a panel of human uveal melanoma xenografts derived from primary and/or metastatic tumors. Clin Cancer Res 16(8):2352–2362. doi:10.1158/1078-0432.CCR-09-3066

Niederkorn JY (1984) Enucleation in consort with immunologic impairment promotes metastasis of intraocular melanomas in mice. Invest Ophthalmol Vis Sci 25(9):1080–1086

Rem AI, Oosterhuis JA, Keunen JE, Journee-De Korver HG (2004) Transscleral thermotherapy with laser-induced and conductive heating in hamster Greene melanoma. Melanoma Res 14(5):409–414

Robanus-Maandag E, Dekker M, van der Valk M, Carrozza ML, Jeanny JC, Dannenberg JH, Berns A, te Riele H (1998). p107 is a suppressor of retinoblastoma development in pRb-deficient mice. Genes Dev 12:1599–1609

Romanowska-Dixon B, Urbanska K, Elas M, Pajak S, Zygulska-Mach H, Miodonski A (2001) Angiomorphology of the pigmented Bomirski melanoma growing in hamster eye. Ann Anat 183(6):559–565. doi:10.1016/S0940-9602(01)80069-8

Romer TJ, van Delft JL, de Wolff-Rouendaal D, Jager MJ (1992) Hamster Greene melanoma implanted in the anterior chamber of a rabbit eye: a reliable tumor model? Ophthalmic Res 24(2):119–124

Rous P, Murphy JB (1912) The behavior of chicken sarcoma implanted in the developing embryo. J Exp Med 15(2):119–132

Schatton T, Murphy GF, Frank NY, Yamaura K, Waaga-Gasser AM, Gasser M, Zhan Q, Jordan S, Duncan LM, Weishaupt C, Fuhlbrigge RC, Kupper TS, Sayegh MH, Frank MH (2008) Identification of cells initiating human melanomas. Nature 451(7176):345–349. doi:10.1038/nature06489

Schuitmaker JJ, Vrensen GF, van Delft JL, de Wolff-Rouendaal D, Dubbelman TM, de Wolf A (1991) Morphologic effects of bacteriochlorin a and light in vivo on intraocular melanoma. Invest Ophthalmol Vis Sci 32(10):2683–2688

Shah AA, Bourne TD, Murali R (2013) BAP1 protein loss by immunohistochemistry: a potentially useful tool for prognostic prediction in patients with uveal melanoma. Pathology 45(7):651–656.

Shields CL, Shields JA (2004) Diagnosis and management of retinoblastoma. Cancer Control 11(5):317–327

Shields CL, Markowitz JS, Belinsky I, Schwartzstein H, George NS, Lally SE, Mashayekhi A, Shields JA (2011) Conjunctival melanoma: outcomes based on tumor origin in 382 consecutive cases. Ophthalmology 118(2):389–395

Singh AD, Topham A (2003) Incidence of uveal melanoma in the United States: 1973–1997. Ophthalmology 110(5):956–61

Sisley K, Rennie IG, Parsons MA, Jacques R, Hammond DW, Bell SM, Potter AM, Rees RC (1997) Abnormalities of chromosomes 3 and 8 in posterior uveal melanoma correlate with prognosis. Genes Chromosom Cancer 19(1):22–28

Smith JR, Falkenhagen KM, Coupland SE, Chipps TJ, Rosenbaum JT, Braziel RM (2007) Malignant B cells from patients with primary central nervous system lymphoma express stromal cell-derived factor-1. Am J Clin Pathol 127(4):633–641.

Spendlove HE, Damato BE, Humphreys J, Barker KT, Hiscott PS, Houlston RS (2004) BRAF mutations are detectable in conjunctival but not uveal melanomas. Melanoma Res 14(6):449–452

Surriga O, Rajasekhar VK, Ambrosini G, Dogan Y, Huang R, Schwartz GK (2013) Crizotinib, a c-Met inhibitor, prevents metastasis in a metastatic uveal melanoma model. Mol Cancer Ther 12(12):2817–2826

Theriault BL, Dimaras H, Gallie BL, Corson TW (2014) The genomic landscape of retinoblastoma: a review. Clin Exp Ophthalmol 42(1):33–52

Tolleson WH, Doss JC, Latendresse J, Warbritton AR, Melchior WB, Jr., Chin L, Dubielzig RR, Albert DM (2005) Spontaneous uveal amelanotic melanoma in transgenic Tyr-RAS+ Ink4a/Arf-/- mice. Arch Ophthalmol 123(8):1088–1094

van den Bosch T, van Beek JG, Vaarwater J, Verdijk RM, Naus NC, Paridaens D, de Klein A, Kilic E (2012) Higher percentage of FISH-determined monosomy 3 and 8q amplification in uveal melanoma cells relate to poor patient prognosis. Invest Ophthalmol Vis Sci 53(6):2668–2674. doi:10.1167/iovs.11-8697

van der Ent W, Burrello C, Teunisse AF, Ksander BR, van der Velden PA, Jager MJ, Jochemsen AG, Snaar-Jagalska BE (2014) Modeling of human uveal melanoma in zebrafish xenograft embryos. Invest Ophthalmol Vis Sci 55(10):6612–6622. doi:10.1167/iovs.14-15202

Van der Ent W, Burrelllo C, de Lange MJ, van der Velden PA, Jochemsen AG, Jager MJ, Snaar-Jagalska BE (2015). Embryonic zebrafish: different phenotypes after injection of human uveal melanoma cells. Ocular Oncol Pathol 1:170–181. doi:10.1159/000370159

Van Essen TH, van Pelt SI, Versluis M, Bronkhorst IH, van Duinen SG, Marinkovic M, Kroes WG, Ruivenkamp CA, Shukla S, de Klein A, Kilic E, Harbour JW, Luyten GP, van der Velden PA, Verdijk RM, Jager MJ (2014) Prognostic parameters in uveal melanoma and their association with BAP1 expression. Br J Ophthalmol 98(12):1738–1743. doi:10.1136/bjophthalmol-2014-305047

Van Raamsdonk CD, Bezrookove V, Green G, Bauer J, Gaugler L, O'Brien JM, Simpson EM, Barsh GS, Bastian BC (2009) Frequent somatic mutations of GNAQ in uveal melanoma and blue naevi. Nature 457(7229):599–602. doi:10.1038/nature07586

Van Raamsdonk CD, Griewank KG, Crosby MB, Garrido MC, Vemula S, Wiesner T, Obenauf AC, Wackernagel W, Green G, Bouvier N, Sozen MM, Baimukanova G, Roy R, Heguy A, Dolgalev I, Khanin R, Busam K, Speicher MR, O'Brien J, Bastian BC (2010) Mutations in GNA11 in uveal melanoma. N Engl J Med 363(23):2191–2199. doi:10.1056/NEJMoa1000584

Versluis M, de Lange MJ, van Pelt SI, Ruivenkamp CA, Kroes WG, Cao J, Jager MJ, Luyten GP, van der Velden PA (2015) Digital PCR validates 8q dosage as prognostic tool in uveal melanoma. PLos One 10(3):e0116371. doi:10.1371/journal.pone.0116371

Weber JL, Smalley KS, Sondak VK, Gibney GT (2013) Conjunctival melanomas harbor BRAF and NRAS mutations–letter. Clin Cancer Res 19(22):6329–6330. doi:10.1158/1078-0432.CCR-13-2007

White L, Meyer PR, Benedict WF (1984) Establishment and characterization of a human T-cell leukemia line (LALW-2) in nude mice. J Natl Cancer Inst 72(5):1029–1038

White L, Trickett A, Norris MD, Tobias V, Sosula L, Marshall GM, Stewart BW (1990) Heterotransplantation of human lymphoid neoplasms using a nude mouse intraocular xenograft model. Cancer Res 50(10):3078–3086

Wilson BJ, Schatton T, Zhan Q, Gasser M, Ma J, Saab KR, Schanche R, Waaga-Gasser AM, Gold JS, Huang Q, Murphy GF, Frank MH, Frank NY (2011) ABCB5 identifies a therapy-refractory tumor cell population in colorectal cancer patients. Cancer Res 71(15):5307–5316. doi:10.1158/0008-5472.CAN-11-0221

Windle JJ, Albert DM, O'Brien JM, Marcus DM, Disteche CM, Bernards R, Mellon PL (1990). Retinoblastoma in transgenic mice. Nature 343:665–669

Yang H, Jager MJ, Grossniklaus HE (2010) Bevacizumab suppression of establishment of micrometastases in experimental ocular melanoma. Invest Ophthalmol Vis Sci 6:2835–2842. doi:10.1167/iovs.09-4755

Yang H, Cao J, Grossniklaus HE (2015) Uveal melanoma metastasis models. Ocular Oncol Pathol 1:151–160. doi:10.1159/000370153

Commentary

Arun D. Singh
e-mail: SINGHA@ccf.org
Department of Ophthalmic Oncology,
Cole Eye Institute,
Cleveland Clinic,
Cleveland, OH, USA

The field of ophthalmic oncology includes a large variety of ocular and periocular malignant tumors. Because these tumors are rare, many aspects of their biologic behavior and possible therapeutic interventions can only be explored using animal models. As elegantly summarized in this review, several different animal models that are spontaneous, transgenic or induced, exist for ophthalmic malignant tumors in mice, rats, chick embryos, zebrafish, and *Drosophila*. Animal models for conjunctival melanoma, uveal melanoma, retinoblastoma, and vitreoretinal lymphoma, among others are under active investigation. With regards Animal Models of Ocular 10 Tumors to uveal melanoma, use of animal models led to determination of optimal settings for transpupillary thermotherapy. Additional uveal melanoma models involving anterior or posterior chamber inoculations have been used to study metastatic behavior, tumor angiogenesis, and tumor immunology. With zebrafish model, effects of drugs on melanoma cell migration has been studied. Similarly, with a new model of retinoblastoma xenograft (injected into the subretinal space), efficacy of intravitreal injections of melphalan and carboplatin are under investigation as a step prior to clinical evaluation. Hopefully, detailed study of animal models will eventually improve the outcomes of patients with rare ophthalmic malignant tumors.

Epilogue

Charles E. Egwuagu

Experimental animal models are the bedrock of modern biomedical research. They provide unparalleled insights into basic pathophysiological mechanisms of human diseases and have been invaluable in identifying, developing, and validating new drugs and treatment modalities. This truism is applicable to vision research as reflected by the chapters in this compendium on animal models of a wide variety of important ocular diseases including cataract, diabetic retinopathy, retinal degeneration, glaucoma, age-related macular degeneration (AMD), as well as infectious and autoimmune ocular diseases. As aptly stated by West-Mays, Bowman, and Yizhi Liu in the chapter on animal models of primary and secondary cataracts, "elucidation of the mechanisms underlying cataractogenesis has only been possible through the use of experimental animal models." Although a variety of rodents (mice, rats, and guinea pigs), mammals (rabbits, dogs, and primates), invertebrates (fruit flies and nematodes), and zebrafish models have been developed for the study of human ocular diseases, the mouse species is by far the most common. Mouse models have gained wide popularity in vision research because the mouse eye is structurally similar to the human eye, manifests many ocular disorders of humans, and has an accelerated life span that allows for investigations of the natural history and progression of eye diseases over relatively short time periods.

Human and mouse genomes exhibit remarkable DNA sequence homology in their coding regions and chapters in this volume have accordingly highlighted the contributions of human homologs of mutated mouse genes to our current understanding of mechanisms that underlie the development of a number of idiopathic ocular diseases. The $rd10$ ($Pde6b^{rd10}$) mouse strain described by Chang is an exemplar of the value of mouse models to our understanding of inheritable eye diseases such as retinitis pigmentosa (RP), a group of inherited retinal conditions that lead to visual field loss and retinal degeneration. Moreover, linkage analysis and genome-wide association studies (GWAS) are often validated in the mouse and human homologs of mutated mouse genes and aid in identifying clinically relevant genes in complex human ocular diseases such as glaucoma and age-related macular degeneration (see Johnson et al.; Landowski et al., Chap. 4). On the other hand, the chapter by Hendricks et al. on mouse models of herpes simplex virus 1 (HSV-1) underscores the critical role of animal models in understanding immunological mechanisms that prevent establishment of HSV-1 latency and mitigate herpetic stromal keratitis (HSK), the leading infectious cause of blindness in the world. As reiterated by Edward Holland, "From a diagnostic and therapeutic perspective, HSV keratitis is one of the most challenging entities confronting the clinician." and "animal mod-

C. E. Egwuagu (✉)
Molecular Immunology Section, Laboratory of Immunology, National Eye Institute, National Institutes of Health (NIH), Bethesda, Maryland, USA
e-mail: EgwuaguC@NEI.NIH.GOV

© Springer International Publishing Switzerland 2016
C.-C. Chan (ed.), *Animal Models of Ophthalmic Diseases*, Essentials in Ophthalmology,

els of HSV are critical to the understanding of the pathophysiology and efficacy of new treatment."

Advances in transgenic technology and bioengineering such as the CRISPR/Cas9 system have ushered a new era in medical research. These recent developments now permit the creation of novel mouse strains for the study of complex multifactorial disorders such as autoimmune diseases, cancer, AMD, and other neurodegenerative diseases. The chapters on transgenic mouse models of diabetic retinopathy (Stitt et al.), AMD and subretinal inflammation (Sennlaub), and orbital and intraocular inflammation (Banga et al.; Kielczewski et al.) illustrate the importance of transgenic mice in vision research and provide the framework for future development of targeted therapies for ocular diseases. As noted by Sheldon Miller in the Foreword to this book, targeted modification of specific genetic loci in transgenic mice by CRISPR/Cas9 has been exploited for the generation of novel mouse models that recapitulate human ocular disease phenotypes, providing unprecedented opportunities for vision research.

Collectively, the animal models described in this book have contributed significantly to our current understanding of the vertebrate visual systems and they will continue to play important role in identifying and testing drug candidates for human clinical trials. While on balance, these animal models reflect essential pathophysiological mechanisms of their corresponding human diseases; doubts have been raised as to whether this axiom is true for complex human diseases such as cancer, autoimmune, and neurodegenerative diseases. In a recent report comparing the genomic response between human inflammatory diseases and murine models, it was found that "although acute inflammatory stresses from different etiologies result in highly similar genomic responses in humans, the responses in corresponding mouse models correlate poorly with the human conditions and also, one another" (PNAS, 110: 3507–3512). This and other sobering conclusions reached by other recent systematic studies that have evaluated how well murine models mimic human diseases raise a cautionary note in context of opportunity cost on our tacit acceptance and reliance on inferences based on animal experimentation. Thus, high priority should be given to evaluating how well our current animal models mimic and translate directly to complex human ocular diseases. As disease in patients with diverse genetic background is much more complex than studying model systems in inbred congenic mouse strains, it is also imperative that causal associations made in animal models be validated in outbred strains that may better reflect the heterogeneity of human populations.

References

Seok J, Warren HS, Cuenca AG, Mindrinos MN, Baker HV, Xu W, Richards DR, McDonald-Smith GP, Gao H, Hennessy L, Finnerty CC, López CM, Honari S, Moore EE, Minei JP, Cuschieri J, Bankey PE, Johnson JL, Sperry J, Nathens AB, Billiar TR, West MA, Jeschke MG, Klein MB, Gamelli RL, Gibran NS, Brownstein BH, Miller-Graziano C, Calvano SE, Mason PH, Cobb JP, Rahme LG, Lowry SF, Maier RV, Moldawer LL, Herndon DN, Davis RW, Xiao W, Tompkins RG; Inflammation and Host Response to Injury, Large Scale Collaborative Research Program (2013) Genomic responses in mouse models poorly mimic human inflammatory diseases. Proc Natl Acad Sci U S A. 110:3507–12

Index

© Springer International Publishing Switzerland 2016   143
C.-C. Chan (ed.), *Animal Models of Ophthalmic Diseases,* Essentials in Ophthalmology,

Printed in the United States
By Bookmasters